ARTILLERY OF THE WORLD

BRASSEY'S
ARTILLERY OF THE WORLD

2nd Edition fully revised and updated

Guns, howitzers, mortars, guided weapons, rockets and ancillary equipment in service with the regular and reserve forces of all nations

Editor and chief consultant: Brigadier Shelford Bidwell, O.B.E.

Co-authors, compilers and consultants: Brian Blunt, Mark Hewish, Ian V. Hogg, Tolley Taylor

Co-ordinating editor: John Buchanan-Brown

BRASSEY'S PUBLISHERS LTD

a member of the Pergamon Group

OXFORD · NEW YORK · TORONTO · SYDNEY · PARIS · FRANKFURT

U.K.	Brassey's Publishers Limited, a member of the Pergamon Group, Headington Hill Hall, Oxford OX3 0BW, England
U.S.A.	Pergamon Press Inc., Maxwell House, Fairview Park, Elmsford, New York 10523, U.S.A.
CANADA	Pergamon Press Canada Ltd., Suite 104, 150 Consumers Rd., Willowdale, Ontario M2J 1P9, Canada
AUSTRALIA	Pergamon Press (Aust.) Pty. Ltd., P.O. Box 544, Potts Point, N.S.W. 2011, Australia
FRANCE	Pergamon Press SARL, 24 rue des Ecoles, 75240 Paris, Cedex 05, France
FEDERAL REPUBLIC OF GERMANY	Pergamon Press GmbH, 6242 Kronberg-Taunus, Hammerweg 6, Federal Republic of Germany

First edition 1977

Second edition 1981

British Library Cataloguing in Publication Data
Brassey's artillery of the world. - 2nd ed., fully rev. and updated.
1. Artillery - Handbooks, manuals, etc.
I. Bidwell, Shelford II. Hewish, Mark
III. Hogg, Ian V.
623.4'1'0212 UF48

ISBN 0-08-027035-2

Library of Congress Catalog Card No.: 80-42161

Printed in Great Britain by A. Wheaton & Co. Ltd., Exeter

Foreword

By General Sir Harry Tuzo GCB OBE MC MA
Master Gunner, St James's Park.

In his foreword to the first edition of this remarkable work my predecessor as Master Gunner referred to the Russian artillery as 'probably the strongest in the world'. Now, writing only three years later, I would delete the word 'probably'. In the light of recent events it is inevitable that the fighting arm to which the Soviet Union gives primacy on the battlefield should occupy an important section of this second edition and should be the object of intense study by those whose interests, amateur or professional, lie in the field of military tactics and the strategies which govern them.

The arguments, which rightly persist, about the nature and effects of artillery fire in its broadest sense are provided here with a corpus of raw material which is as authoritative as it is readable. Opinions on the merits of guns as against rockets, on indirect fire as against flat trajectory, on methods of guidance and fire direction and, perhaps above all, on the best way to deal with massed formations of tanks—these issues and others like them can start from the firm base of facts given in this book. Perhaps, also, the trend towards stronger conventional deterrence will be assisted by the information given on the remarkable accuracies now attainable both on the surface and in air defence.

The book, in all ways a worthy successor to the first edition, leaves the reader in no doubt as to the capabilities and diversity of Warsaw Pact artillery; but, as its name implies, the scope of the work is world-wide and we are again presented with the meticulous research, clarity of exposition and sheer comprehensiveness which marked the earlier volume.

The obvious necessity to include new and projected equipment has led to the deletion of a few weapon systems which may still be in service in remote areas of the world but are, none the less, technically obsolete. The great 'French 75' is one example. I believe that few will quarrel with this shrewdly exercised selectivity when it is considered what we gain. We have here a book which is leaner and more compact than its forerunner but still amply reflects the knowledge and great attention to detail of its highly skilled compilers.

I am grateful for the opportunity to write this foreword and thus to be associated with a work which is outstanding in its class and which meets an important requirement in the world of military studies.

Acknowledgements

The task of compiling this inventory of artillery weapons in service and of collecting illustrations was only made possible by the generous cooperation and assistance we have received from various official sources, from firms and from individuals. It must be stressed that the information on which it is based is from open sources available to the public: where officials and firms manufacturing defence equipment have given invaluable help is in guidance and indicating where to look. The use we have made of the information and the responsibility for errors or omissions is ours alone.

We would like to record our gratitude to the many firms consulted, and to the representatives of Defence Ministries of the Federal Republic of Germany, France, Israel, Sweden, Switzerland, the United Kingdom and the United States. We have had particular help from the United States Field Artillery School, Fort Sill; the Royal School of Artillery, Larkhill; the Royal Artillery Institution; the Imperial War Museum; The International Institute for Strategic Studies; J. F. Lehmanns Verlag: and we owe special thanks to Brigadier Michael Calvert; Brigadier J. C. Codner CBE MC, Editor *Journal of the Royal Artillery*; Major C. J. Davidson of Forenade Fabriksverken; Colonel P. R. R. de Burgh OBE, Librarian the Badley Library, Royal School of Artillery; Major Robert Elliott, International Institute for Strategic Studies; Captain B. L. Grawe; Lieutenant Colonel M. L. O'Hagan RA (by whose permission we have based our description of BATES upon his article in the *Journal of the Royal Artillery* Vol CVII No 2); and Mr Jac Weller.

Drawings: Eric Rose and Pierre-Andre Tilley of Blitz Publications

Contents

Introduction

The aim of this book is twofold. It provides in a convenient form a catalogue of all artillery systems likely to be found in service with the field armies in the world today as complete in detail as possible when relying only on overt sources and published information. It also explains in non-technical terms the role of the various categories of gun and rocket, and their method of employment in the major armies of the world.

Artillery weapon-systems are technologically complex, and properly operate from outside the actual combat zone. Accordingly, they tend to be overlooked or ignored in defence studies, where the 'guns' are overshadowed by the far more dramatic tanks in the ground battle, by strike aircraft and, more recently, anti-tank and guided missiles. It is often forgotten that successful ground operations in conventional war are the result of close coordination between infantry, armour and artillery. At least half the ground-based fire power on the battlefield is provided by artillery, and the gunner's vital contribution is unaffected by climate, weather or time of day: it is also less vulnerable to enemy interference. The weapon inventories of the Soviet and Warsaw Pact armies reflect a much greater understanding of this fact than is apparent in the NATO order of battle. It is hoped therefore that this book provides information essential for the military man, the defence orientated general reader, the military commentator and the defence correspondent.

The term 'artillery' is a wide one which can be applied to a whole spectrum of weapons ranging from the simplest cannon designed for use at short ranges over open sights, i.e. with the target in full view of the gun's crew, to rocket-propelled guided missiles with computerised fire-control systems. The difficulty is to decide on boundaries between categories. Strictly speaking, artillerymen regard all 'cannon' as lying within their province, provided that they satisfy the criteria of employing sophisticated delivery systems, e.g. either indirect fire or guidance, are crew-served and have range and lethality exceeding those weapons found normally in the tables of equipment of infantry or armoured units. This leads us to include all tactical weapons fired from stationary land platforms and to exclude some hand-held, man-portable weapons, the weapons in armed helicopters and the cannon in armoured fighting vehicles, including tanks, although these may well be in some cases larger and heavier than some systems which are by our definition 'artillery'. A wide range of indirect fire artillery still in service, however, has a secondary anti-tank direct fire capability, and some armies still retain classic type towed field guns designed specifically for anti-tank defence; these are included as they fall into the general category of cannon. The family of air defence weapons includes both surface to air guided missiles and guns. Guns proved to be as valuable as ever in the Arab–Israeli war of October, 1973.

The term 'indirect fire' crops up so regularly and is so frequently misunderstood by the non-artilleryman that it must be explained here: else much of what follows will not be clear. *Direct fire* is extremely efficient and effective, using as it does the best target acquisition system over short ranges yet designed—the human eye, and the cheapest computer—the human brain. It is simple to operate; a vital consideration for hastily trained conscript armies or part-time national armies such as Israel's. It also permits the cannon to be used at its optimum accuracy, say 1,000–2,500 metres, where the chance of a first round hit is very high. But all direct fire guns are exposed to equally effective counter-fire: therefore the crew must be protected and we end up with the tank. As the range is short, the area dominated by a single tank is very small and so large numbers of tanks are required for a given mission. A gun with a range of 15/20 kilometres dominates an enormous area by comparison.

Indirect fire arose from the need to conceal long-range artillery systems from counter-fire. The weapon therefore became a *weapon-system*, made up of the gun itself, a means of acquiring targets on the now distant battlefield, a network of communications and a technical and tactical fire control system.

Immutable laws of physics dictate that the accuracy of a ballistic projectile decreases with range. To reduce the limitations imposed by an inherent spread in the fall of shot, or 'zone', the concentration of as many guns as possible on to the target is an unchanging artillery maxim: thus the battery and not the individual gun is the basic fire unit.

By correcting the fall of shot, an observer overcomes, or 'shoots out', all the inaccuracies in the weapon-system except the zone of the gun. This arrangement seems ideal except that it is slow because of the time taken for each round to fly from gun to target. Furthermore, the system of observed indirect fire depends on having a competent observer in the right place, and his seeing the target.

The basic data needed to set the gun sights for indirect fire is calculated from the known coordinates of gun and target positions. The processes of determining these coordinates are similar to those used in land survey, except that they have to be done with great rapidity. The meteorological data must be continuously measured, the muzzle velocity of each gun known, and so on. Calculations to assess the effect of all these and provide correct data have to be made for every gun and every mission, manually in most systems but by computer in technologically advanced armies.

The only way to reduce delivery error is to use guidance. Delivery error can never be eliminated completely, but for practical purposes the operator and not the mechanism of guidance is the source of guidance error. There are a wide variety of guidance systems, some of which rely on monitoring the trajectory and transmitting correctional data while others use inertial guidance, by which the missile carries its own self-sufficient navigational system, and is therefore invulnerable to electronic counter measures (ECM) or 'jamming'. The latest and most refined technique is terminal guidance, by which a free ballistic trajectory is corrected in the last seconds of flight by homing devices based on lasers or electro-optical systems. These have already been used in aircraft bombs, but the system is adaptable for artillery shells or surface-to-surface missiles. Compared with ordinary free flight shell or shot guidance is immensely expensive in terms of research and development as well as cost per round fired. Conventional artillery is labour intensive, using elementary techniques, and a large number of relatively cheap missiles for suppressive and harassing fire. Guided missiles are not only capital intensive, but highly cost effective, achieving ideally an exchange rate of one missile per mission on target.

Whatever system is used it must be emphasised that the actual vehicle for carrying the lethal agent, whether this is ordinary chemical reaction high explosive or nuclear, is only one element of the weapon system. All artillery systems have an inherent error which we minimise by various means, but the total accuracy also depends on the accuracy of target location. The mathematical relationship is such that one can say without over simplification that money spent on target acquisition and surveillance equipment doubles or even redoubles the effectiveness of the artillery system as a whole. As we shall see when we come to the field artillery section the crude, mass-fire blanketing techniques developed in the period 1917–1945 are being replaced by far more sophisticated methods aimed at maximum effect for minimum shell use.

Before going on to a detailed study of the various categories of artillery there is one basic question which must be answered and kept in mind throughout by military students when considering the subject. What is 'artillery' for? In other words, how does it differ from the other arms, considering that each of them in their own way are producers of fire power?

Indirect fire field and anti-aircraft artillery are essentially deterrents. To be sure, they are designed to have the best probability within the constraints of the system to inflict lethal damage, but they can achieve their object indirectly by forcing the enemy to adopt some other mode of attack or to desist from attacking for fear of an unacceptable casualty rate. For instance, well sited air defence weapons can deter strike aircraft from using the optimum height and angle of approach; medium artillery of 152 mm or 155 mm calibre can take the sting out of a tank attack by damaging the tanks themselves or by causing casualties among the accompanying motorised infantry and SP artillery; artillery covering fire can suppress the fire of troops in defensive positions by direct attack, or by blotting out their observation with smoke screens. Unguided rocket systems are used to obtain the same effect, saturating the target with large numbers of ripple fired missiles which arrive in the target area within a few seconds. Guided weapons are essentially different, in that an expensive system is used to carry a large or at any rate, a highly lethal warhead to a high-value target and the aim is not to 'neutralise', to use the British tactical term, or render temporarily ineffective, but to destroy.

The same is true of the out-dated but still effective categories of anti-tank and anti-ship guns. A near-miss may inflict superficial damage, but is still a miss; to be effective, the target must be actually struck and set on fire or sunk. Anti-tank guns rely on the simplest and most primitive methods. Coast defence artillery, now virtually obsolete, is as sophisticated as anti-aircraft artillery, requiring radar and computerised fire control.

To summarise, we therefore consider in this book six categories of artillery equipment, each prefaced by a note on tactical employment and the main features of design and operation and two on ancillary equipment:

(a) Field artillery—guns and howitzers;
(b) Mortars;
(c) Battlefield surface to surface missiles carrying either nuclear or conventional warheads;
(d) Ancillary equipment essential to surface-to-surface systems—fire-control, surveillance, target acquisition, survey and for providing ballistic data.
(e) Anti-tank guns and long range anti-tank guided missiles;
(f) Air Defence artillery;
(g) Ancillary equipment essential to air defence systems: surveillance and tracking radars, and fire control computers;
(h) Coast artillery, or 'anti-ship' artillery.

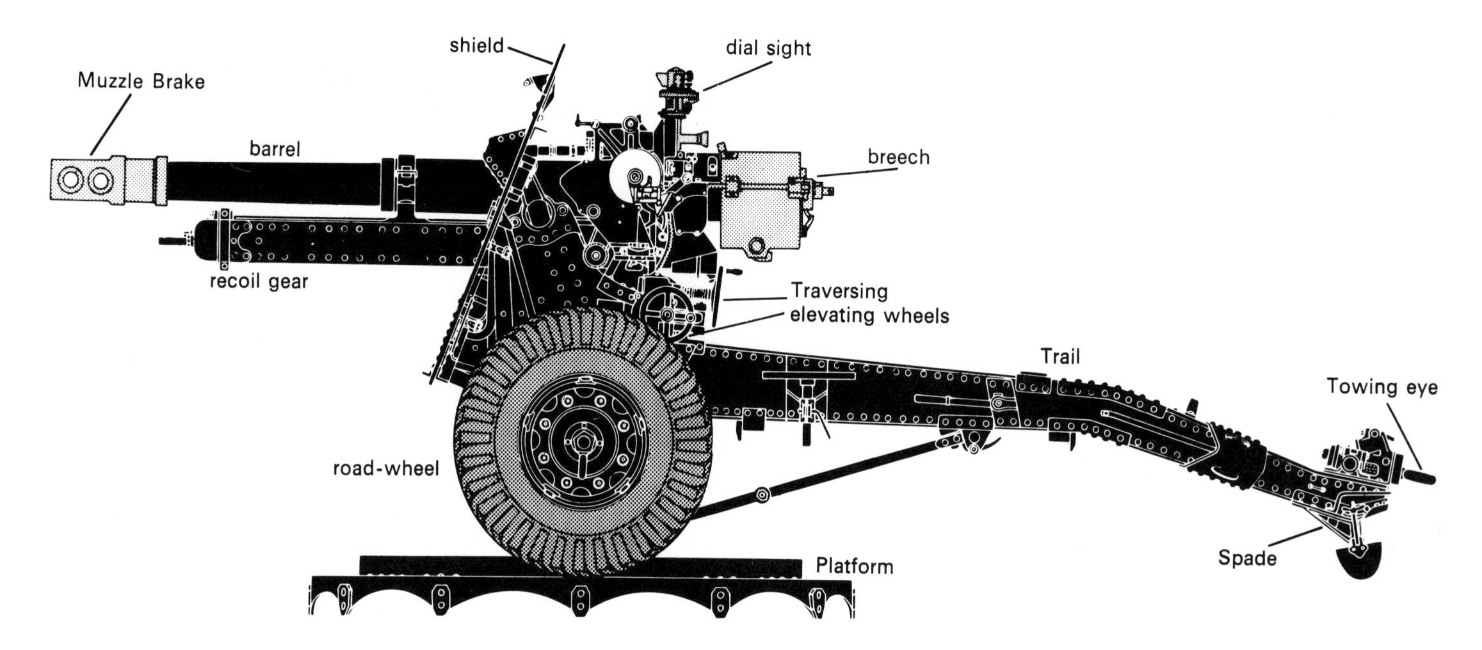

Figure I: Anatomy of the Gun

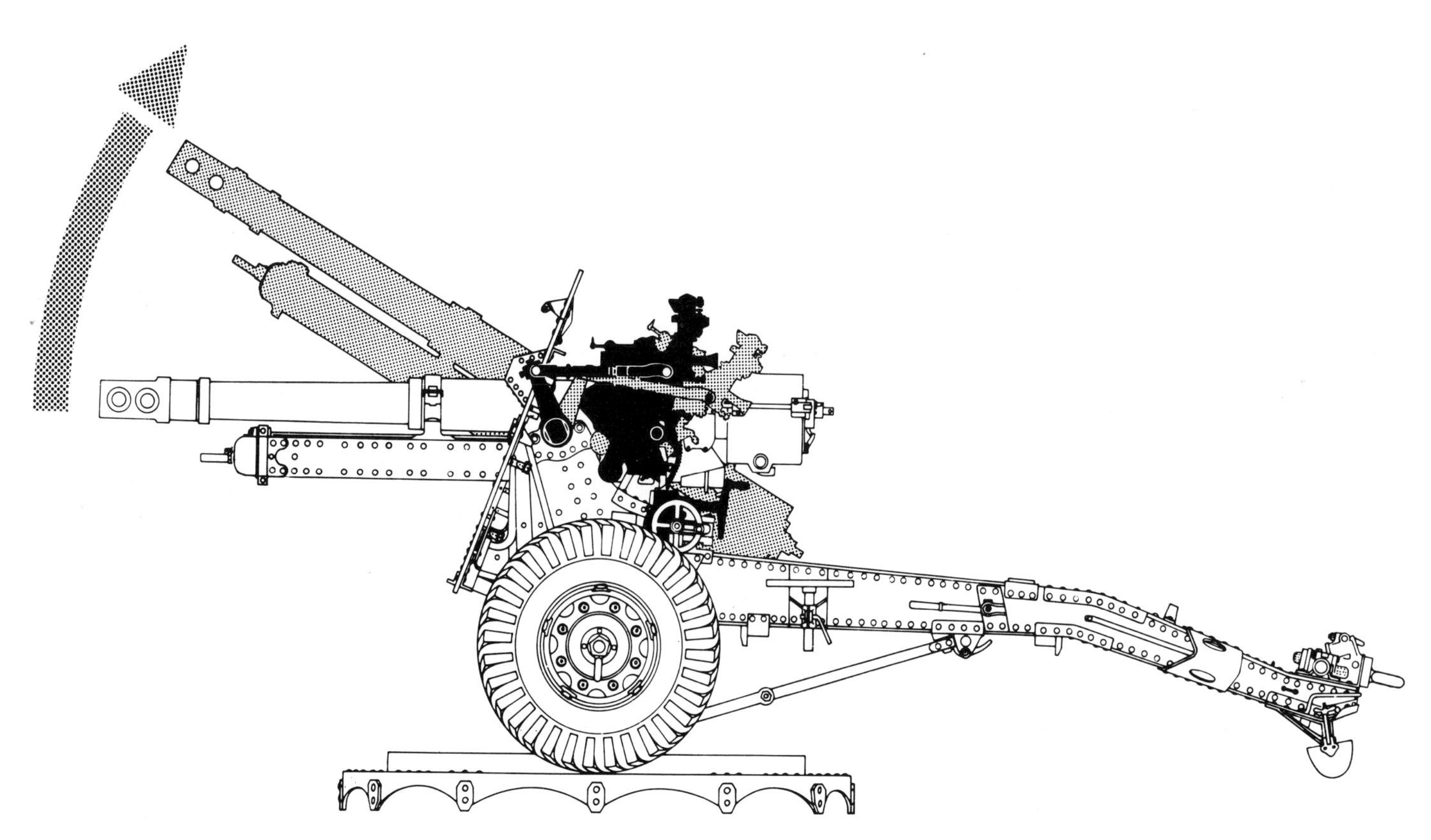

Figure I(a): Elevation

I. Characteristics of Guns and Howitzers

The following remarks on gun design and characteristics apply equally to field artillery, and the guns covered by the other sections in the book. The general comments will not be repeated in the introduction of each section, except where they are of singular importance.

The basic layout of a gun is determined by its primary role and any constraints imposed by the user. The requirement for a field gun will state the desired range and the target effect. Special constraints may include a weight limit or the desire to take advantage of existing ammunition types. The user will also dictate the priorities between the inevitable design compromises: thus the barrel of the Italian Pack Howitzer is kept light enough for pack transports, accepts an existing ammunition (American M 1) but has too short a range for modern tactical needs.

Such diverse requirements make a statistical comparison between the efficiency of different weapons difficult. A quick assessment of performance is given by the equation:

$$\frac{\text{Shell Weight} \times \text{Range}}{\text{Equipment Weight}}$$

This equation shows that the Russian 152 mm D 1 is 30 per cent more efficient than the older 152 mm ML 20, which is fair comment on ten years of development.

A gun consists of a number of components as figure 1 demonstrates. For convenience these components are divided into two 'sub-assemblies': the recoiling parts, and the mounting. The recoiling parts consist of the barrel (sometimes referred to as the piece) muzzle brake, jacket, breech ring, breech block and moving parts of the recoil system. The mounting consists of the rest of the equipment, and may be a wheeled carriage, a complete vehicle, or a fixed emplacement. The term ordnance is often used to describe a weapon, but is properly the recoiling parts less any part of the recoil system.

The gun barrel must absorb many stresses. There are three main types: the great pressures of the explosive propellant gas presses outwards against the walls and is greatest in the area of the chamber: the shell being twisted by the grooves in the rifling, tends to stretch the barrel lengthways: the barrel must be strong enough to support its own weight without drooping at the muzzle. The size and weight of the barrel structure, its wall thickness and jacket (if fitted) will be dictated by the strength of the steel used. If the barrel can be pre-stressed it will be stronger for a given size and quality of steel.

One method of construction is to build-up the barrel from a number of concentric tubes, which are shrunk onto each other and are thus pre-stressed. Most modern barrels are of monobloc construction, that is to say that the barrel is made from a single forging. In 1909 the British suggested that a gun-barrel could be strengthened by subjecting it to intense hydraulic pressure and so shrinking the outer layers of steel over the inner, and in 1913 the French put the idea to practical use when making a 140 mm gun. The process is known as autofrettaging and allows the use of relatively low grade steel while retaining the manufacturing simplicity of a single forging.

Central to the gun's design is the detail of chamber and breech. The options revolve around the choice of ammunition type. It is clearly an advantage if the gun accepts the ammunition of existing equipments and that used by allies. This desirable bonus does not

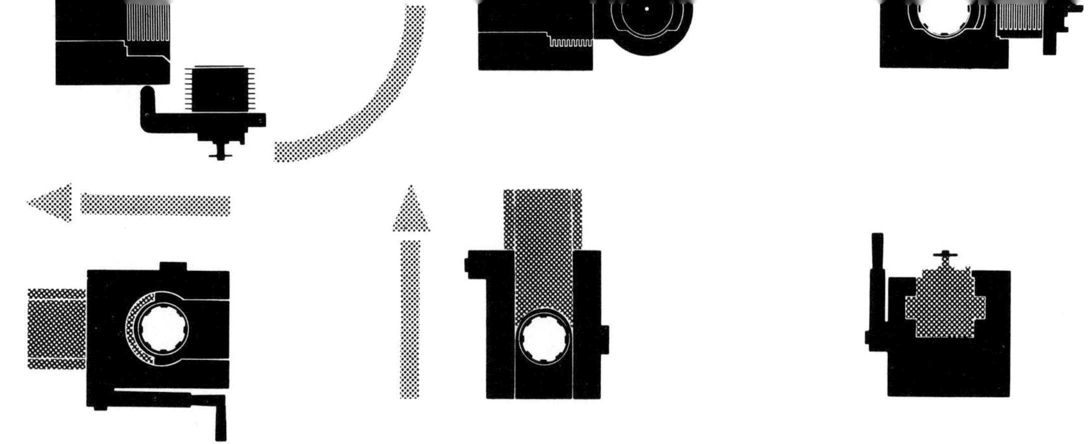

Figure II: Breech Mechanisms

The choice of breech block design is related to the ammunition type, and will affect the rate of fire. The strongest, and oldest of present systems is the interrupted screw type. It was developed by the French at the end of the nineteenth century. Unlike earlier attempts, the threads are not continuous so that the block does not have to be literally 'unscrewed': they run in arcs of 45° with a clear machined space between: a quarter turn of the block disengaged the threads and allows the block to be pulled or swung clear. The threads and machined surface of the block are tapered, so is the corresponding opening in the breech ring: the quarter turn of the closing movement forces the breech inwards to mate the tapering machined surfaces. The design lends itself to good obturation, and is commonly, but not exclusively associated with bag charge ammunition. Its strength leads to its exclusive use with heavy guns.

Above
Example of vertically sliding breech block

Top left
This interrupted screw thread breech is designed for use with a cased charge. The firing pin passes through the hole in the face of the block

Bottom left
Example of horizantally sliding breech block

The alternative is a sliding, or sliding wedge block, moving horizontally or vertically. Its strength depends entirely on the open ended claw in which it moves. The actuating lever moves in one direction only to operate this type of block, and this allows varying degrees of automation. On closing, the block moves in a plane off-set from the block face so that it closes into the face, to ensure the correct seating of the cartridge (and with fixed ammunition the shell) and forces home an obturator pad or ring if bag charges are used.

Between a fully automated breech and a manual one lies the semi-automatic action. A semi-automatic can act in both directions—e.g. the German 88 mm Flak 36 was spring-loaded both to open and to close the breech. The breech opening rarely takes place on recoil, almost always on run-out. 'Semi-automatic' does not define the movement; it points to the fact that loading and closing the breech are governed by human intervention and are not mechanically linked to breech opening and extraction.

The rate of fire is dictated by a number of factors including the weight of shell, size of detachment, access to the breech, breech design and the ammunition type. Few towed guns run to the luxury of automatic loading, because of the bulk of machinery involved: however, the paramount need for a high rate of fire from air defence guns makes the penalty of a large carriage worthwhile.

Easiest to load is fixed ammunition since only one action is needed. This advantage can be combined with a variable charge by the semi-fixed round: the shell and cartridge case are mated (or re-mated) for loading after the charge has been adjusted. The popular American 105 mm M 1 range is of this type. Separate ammunition is a slower option requiring the loading and ramming of the shell, then the loading of the propellant. Power ramming is very desirable, especially for heavier shells: it saves effort and leads to a consistent ram which is vital to the gun's accuracy. Bag charges (which have economic

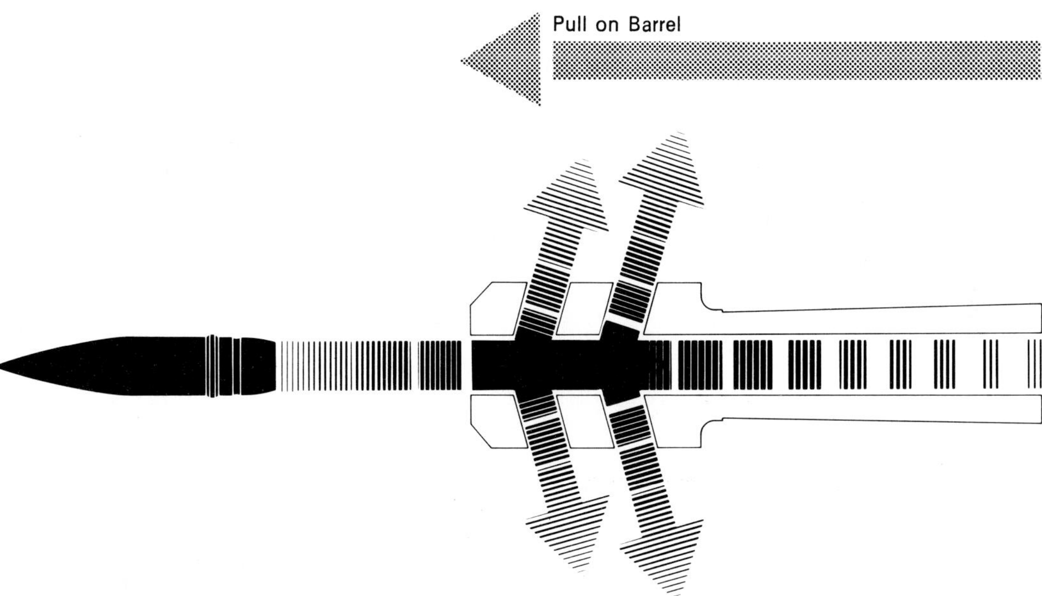

Figure III: Muzzle Brake

and logistic advantages) require a third action, which can be concurrent with the other two, in loading the primer cartridge. A minor inconvenience with bag charges is the need to check that there are no smouldering remnants from the previous charge before a new one is loaded.

An important recognition feature of a gun is the muzzle brake. Amongst the few guns not to have one are the two most widely used American equipments M 114 and M 101. A muzzle brake provides rearward facing baffles or flaps against which the propellant gases react as they move forward out of the bore behind the shell. The effect is partly to counter recoil, and partly to stabilize the gun at the moment of departure of the shell.

The recoil system must absorb the shock of firing—which is the reaction to the enormous acceleration of the shell in the bore—leaving the carriage stable. There are two components to the recoil system—the buffer which stops the recoil, and the recuperator which

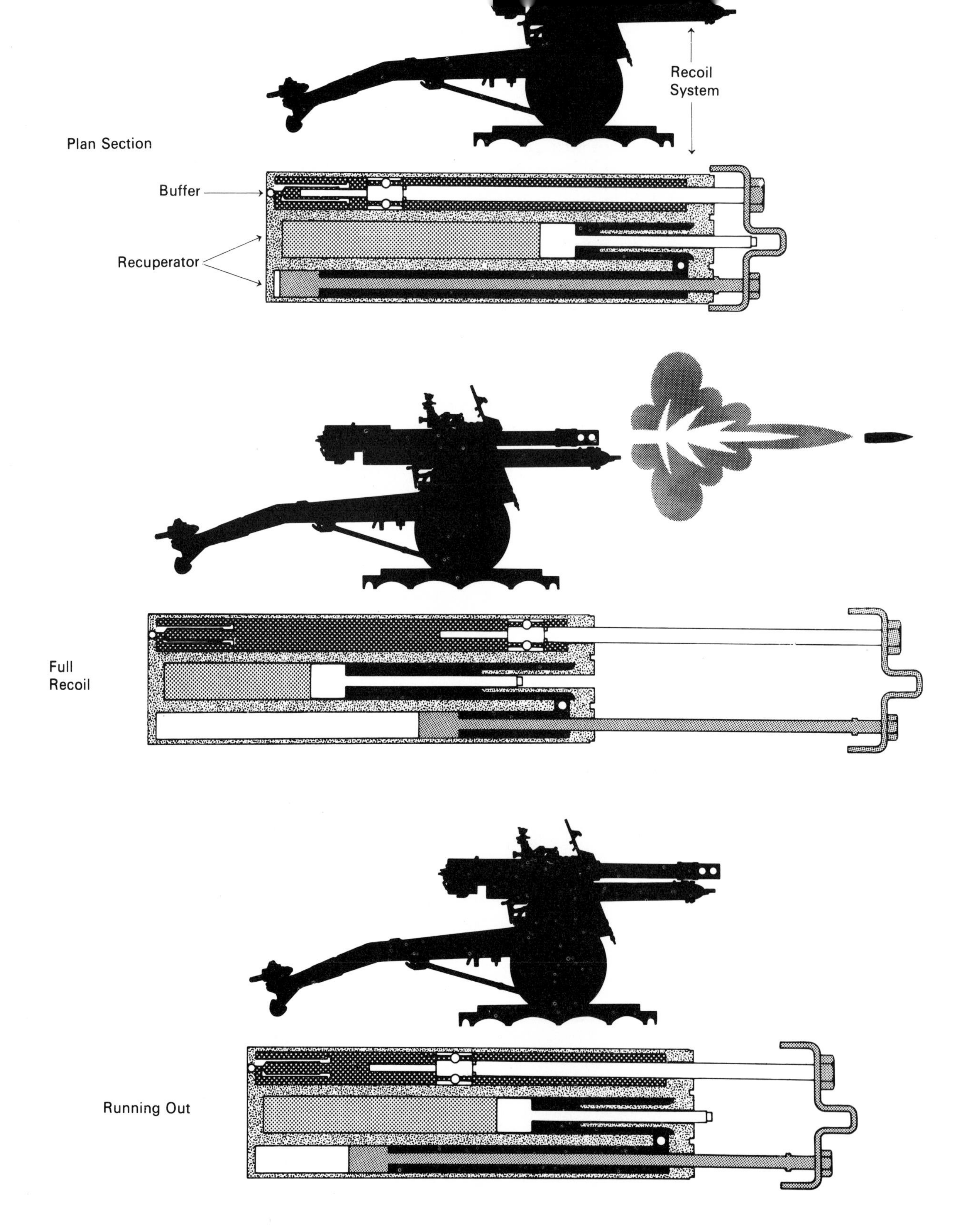

Figure IV: Recoil System of 25 pounder

Note complexity of outwardly simple weapon. Recoil is checked by flow of oil in buffer (dark stipple) through ports in buffer piston. At same time oil in recuperator (black) acting on free piston (white) compresses air (light stipple) which re-expands to cause run out to firing position (below) returns the recoiling parts to the firing position. The buffer is almost invariably hydraulic because this allows for the progressive deceleration of the recoiling mass. An arrangement of mechanical limbs and adjustable valves can be included in the design to shorten the recoil as the elevation increases. This avoids the need for a recoil pit beneath the breech, and the greater stress passed to the carriage as a result is acceptable because it has a greater vertical component and does not therefore, tend to cartwheel the gun. The recuperator is either hydro-pneumatic or a mechanical spring.

Carriage design is influenced by the use to which the weapon will be put. The earliest, pole trail arrangement has now been totally eclipsed because of its limitation on the arc of fire in both azimuth and elevation. The box trail is an improvement, but still limits the traverse available without moving the trail. This arrangement has found favour in the specialist application to the German 90 mm anti-tank gun, but here the need is for ultra-rapid deployment, which this layout permits.

The most common layout is the split trail—two trail legs which are brought together for travelling and spread in action. Traverse of the barrel is limited by the need for horizontal recoil stresses to pass within the arc of the legs. This is an operational limitation which is more apparent than real. Properly handled, batteries can be deployed and redeployed to cover the battle within quite narrow primary arcs of fire. An equipment with a traverse of less than 45° will generally impose an operation limitation or problem. If the gun has to be traversed complete, the difficulty increases with the size and weight of the equipment. The effort of freeing bedded in spades must not be underestimated. A useful expedient is the provision of castor wheels on the trail legs.

Some split-trail carriages are provided with a firing jack attached below the saddle whose purpose is to raise the forward part of the equipment and wheels off the ground, provide three-point support and absorb the vertical component of the recoil force, so making for greater steadiness.

The ability to fire in a 360° arc from a static platform has its attractions. In practice, good gun positions which enable the exploitation of such a potential are few and far between. However, this capability is often seen as worthwhile, although its achievement can lead to other limitations, such as a weight penalty. One solution is to mount the saddle directly on to the centre of a three or four spoked arrangement of trail legs. Another is to use a platform on which the road wheels rest. The American M 102 combines the box trail and platform configuration used by the old British 25 pounder. Because the platform absorbs horizontal recoil forces, the trail can be carried on a longitudinal roller on which it rotates.

The Auxiliary Propulsion Unit (APU) is not a new idea, but one in which interest has recently been re-awakened. Guns like the Swedish FH 77A and tri-national FH70 have been designed to provide limited self propulsion. The system is popular on the small Soviet equipments. The APU provides power for loading and laying as well as assisting deployment. With APU towed guns can occupy platforms inaccessible to heavy tractors and also, if detected by hostile surveillance, redeploy locally without bringing the tractors up from the rear.

The alternative to the classical wheels-plus-trail design is the 'self-propelled' (SP) gun mounted on armoured, tracked chassis for indirect fire. It is an artillery weapon and not to be confused with the tank it so closely resembles externally and which is a direct-fire close-combat vehicle. Developed originally in the 1920s by the British for the close support of tanks, one school of artillery thought considers that now, when all formations are mixed infantry/armour and protection is required against nuclear as well as kinetic weapons, the SP is the only viable equipment for full scale war. There are, however, conflicts of cost, mobility, simplicity and flexibility.

Clearly the development of a simple split trail carriage is quicker and easier than of an SP chassis (or even the conversion of an existing tank design). The production cost per unit is also much smaller. Often, a new barrel can be fitted into an existing carriage, or the same carriage used for different equipments: the Russian M 1938/M 1943 122 mm and 152 mm guns are an example of carriage sharing.

It would be an over-simplification to say that any vehicle will tow a gun, but *in extremis* when battle damage or mechanical failure has immobilised the specialist artillery tractor, this is the case. The towed gun, man handled or hitched to any passing vehicle as an expedient, can still be redeployed. Other aspects of the towed gun's flexibility of movement may depend on its size, but transport options for this type of equipment can include animal packing, portee (i.e. carried in a vehicle), air movement by aircraft or helicopter and parachute delivery. Examples of guns expressly designed to exploit these options will be found in this section.

Simplicity is not just a function of design and construction. It also makes for a more robust equipment, and one which is easier to maintain. It is also important as a function of training. There are real difficulties in teaching and up-dating operating drills in the confines of an SP turret. These are most apparent in a conscript army.

To set against these formidable arguments, there are virtues in the SP gun which find favour in Western armies: it is highly mobile, easier to deploy, can be re-deployed fast and has a greater chance of survival.

The SP has an excellent cross country performance. It enables the gun to travel over damaged roads, round traffic blocks and point-to-point across country. It can be made to swim. This is all highly important in getting the gun to its firing position because a gun on the move is useless. On the battery position, the SP can drive into the best platform (a virtue shared by towed guns with APU) and dispenses with the need for concealed tractor positions. Track planning is easier.

The SP is quicker to bring *into* action than the towed gun because the limiting factor in this drill is the time taken to establish orientation. The SP does score in the immediate availability of ammunition, and the avoidance of double-handling first line holdings off the tractor. The SP is also much faster at coming *out* of action, thus reducing overall travelling time.

Automatic loading is very difficult to achieve with a towed gun, and magazine loading not unheard of. It was incorporated in three now obsolete AA equipments; e.g., the US 75 mm Skysweeper. The SP has the power supply necessary for either or both these functions, with consequent benefits in a high rate of fire and in consistency of loading and target accuracy. The ultimate in present service is the Swedish VK 155 L/50. This is a bulky equipment but achieves a rate of fire of fifteen 155 mm rounds per minute.

Survivability is not an inherent quality of the SP. Indeed, it is vulnerable to loss through mechanical failure. Open designs such as the American M 110 and French Mk III are no better protected than their towed equivalents. However, enclosed types, which make up the majority, offer splinter protection to the crew and delicate components such as sights. In nuclear war, the enclosed SP offers flash protection, and air filtration which also serves to protect the detachment from chemical attack.

In most cases the SP is built on an existing chassis often with modification to lock the suspension for greater stability during firing. However, the SP is not always a hybrid weapon and the Americans are notable for producing 'one-off' chassis: e.g., the M 109 and M 110/107. One disadvantage of the use of a tank chassis is the unnecessary weight of hull armour, the removal of which can negate the cost and development benefits of using an existing design. When a tank chassis is used, it usually travels 'backwards' because whereas the tank engine is behind the turret, access to the breech area of the gun for ammunition supply makes a front engine essential.

Finally a word about sighting and laying. In general, direct laying is only used against moving targets such as tanks and aircraft. The aim has to be offset in azimuth to compensate for the movement of the target while the projectile is in flight: this offset is known as lead and can be set by using graticules in the sight, or by offsetting the whole sight so that the point of aim remains at its centre. The latter system (centre laying) is easier to operate.

Adjustments to elevation to compensate for range can be made in the same two ways using graticule or centre laying. A system using a fixed fore-sight and an adjustable back-sight is known as a tangent sight: the size of the rear sight needed to accommodate large angles of elevation is such that the tangent sight is only suitable for direct laying. The alternative is a rocking bar sight which incorporates both fore- and back-sights in one mounting and is offset by the amount of elevation required and the gun and sight elevated until the sight is realigned.

It will be an advantage if the direct fire sight magnifies the target, but care must be taken in choosing the degree of magnification so that the field of view is not so small as to make it difficult to pick up the target.

Indirect laying in azimuth is achieved by using a 'dial sight', which is a periscopic device free to rotate about a vertical axis. It has a suitably graduated scale. On coming into action the gun is accurately laid in a known bearing, and the dial sight turned to align with a reference object. To lay on a target the sight is turned by an amount equal to the difference between the known bearing of the gun and that of the target, and the gun is traversed until the sight is realigned with the reference object.

The elevating sight is of the rocking bar type but the lay of the gun is determined by a spirit level as opposed to the alignment of the fore and back-sight, as is the case with direct laying. The sight must compensate for any difference in level of the trunnions, and one which does this automatically is called a 'reciprocating' sight.

Reference is made to 'split-function' laying in some of the equipment descriptions: this is an arrangement whereby laying for azimuth and elevation is separated into two functions employing two crew members who usually work one on each side of the barrel.

II. Ammunition

The number of shell and warhead types available to the artilleryman is increasing, and the efficiency of familiar types is improving. Design and development is influenced by the nature of the expected target and by the desired target effect.

The most useful type of shell with the greatest variety of applications and a multiple target effect is High Explosive (HE). It consists of a hollow shell of cast steel fitted with one of a number of chemical explosives.

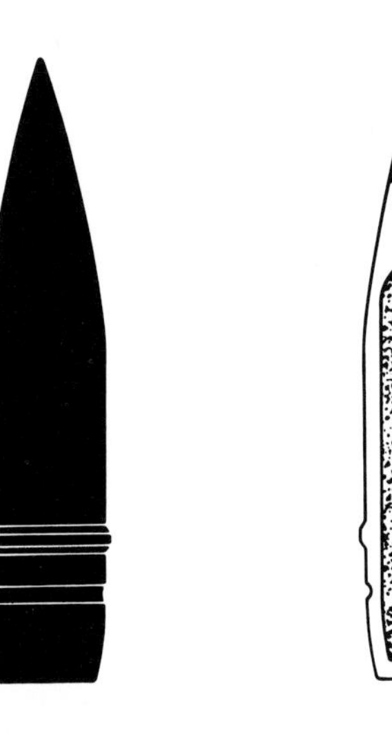

Figure V(a): Basic shape

High Explosive (HE)

A high explosive is a substance which, when subjected to a shock wave of a suitable size and shape, will, in rapid and exothermic reaction, decompose mainly into a gas with a violent rise in pressure and temperature. It must be insensitive so that it can stand the shock of discharge from a weapon, and also (APHE) impact on the target. It should be cheap to make and be safe to manufacture and store. An ideal filling will give the highest possible density, freedom from cavitation, and good internal strength. One of the best methods of filling from the economical aspects is casting at a temperature of about 80°C: higher temperatures are dangerous and lower temperatures give problems in storage and handling in the tropics. Trinitrotoluol (TNT) can be handled in this way and is widely used either on its own or mixed with other explosives. When mixed with ammonium nitrate it is called 'Amatol', and with PETN as 'Pentolite'.

A higher energy explosive was designed by the UK Explosive Research Department in the 1930s and given the name 'Research Department X' or 'RDX'. In its original form it was too positive for use as a projectile filling and was mixed with TNT and beeswax to render it more stable. RDX/TNT mixtures are known as Hexogen. RDX has a high melting point and is sometimes used with a small proportion of inert wax such as paraffin wax and press filled into shells and rockets to achieve a very high density and energy content. A derivative of RDX is HMX which has a greater energy content per cubic centimetre and is now being used in a number of shell fillings, rocket heads, and in HEAT projectiles which require a high performance in a small-calibre, hollow-charge warheads. To increase the blast effects of explosives, aluminium can be added. Aluminium and magnesium can also be used to increase the incendiary effects.

The explosive effect of HE damages both man and material, and the heat generated has an incendiary effect, but the radius of damage of the shell is greatly increased by the flying fragments of the casing. The composition and manufacture of the shell case must enable it to withstand the shock of firing and on detonation, shatter into fragments of a regular size. The optimum size of fragment is large enough to possess sufficient kinetic energy to inflict damage, yet small enough to yield the maximum number of fragments from each shell.

In recent years development has concentrated on pre-fragmented shells: by casting a pattern into the internal surface of the shell case, the size and regularity of the fragments is assumed. Care must be taken not to weaken the shell so that it fails under the stress of firing.

The external shape of the shell with its ogival nose is designed to give the best performance in flight. Given the calibre the weight of the shell is determined by the length of the shell (a function of good ballistic shape and the size of the chamber) and the proportion of explosive filling to metal case content.

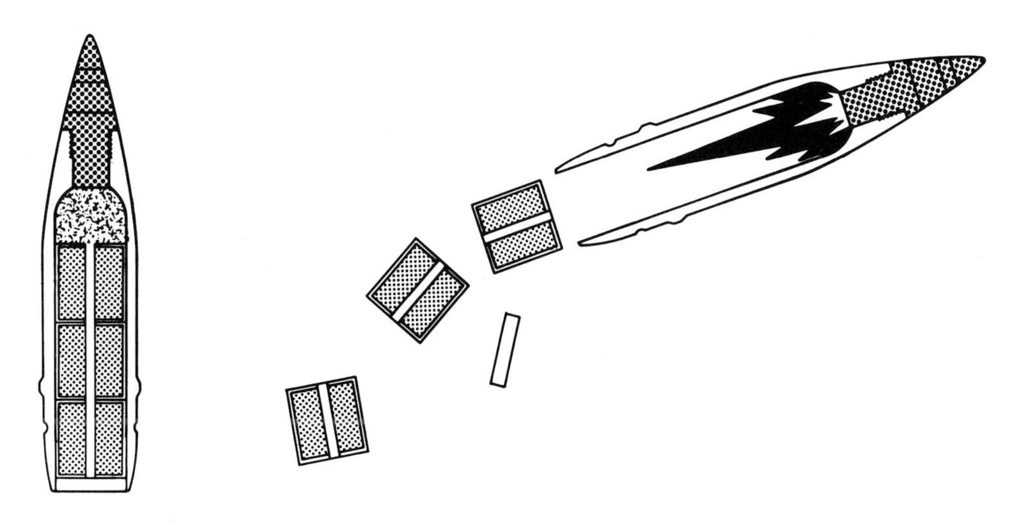

Figure V(b): Carrier shell—Base ejection smoke, illuminating, minelet, chemical, leaflet

In the conventional indirect fire role, the most common artillery shell after HE is smoke. White phosphorus is one sort of smoke agent whose chief virtue is an instant effusion of dense smoke. Its burning action gives a valuable bonus effect against both men and material. Alternatively hexa-chloro-ethane can be used to produce smoke: it burns slowly to give a slower build-up than white phosphorus but a longer lasting effect. Hexa-chloro-ethane is usually delivered in cannisters in a carrier shell whose base is blown off over the target area using a time fuze and small explosive charge. The cannisters are ejected and fall in a random pattern and yield a good spread of obscuration. The two types are known as WP and BE (base ejection). They are not competitive but complement each other: WP's instant effectiveness is offset by its tendency to 'pillar' in vertical columns on still days: BE smoke is less prone to this effect. BE smoke can be coloured in manufacture and used to indicate targets to ground attack aircraft.

Chemical, propaganda (or leaflet) and illuminating shells are all variations on the basic carrier shell. Owing to differences in balance and an altered centre of gravity these projectiles have different ballistic characteristics from the 'standard' HE shell.

Conventional direct fire anti-tank weapons, both guns and guided missiles use specialised armour defeating ammunition. Many field artillery equipments have an anti-armour projectile, useful for the defence of the gun position, but also offering a secondary anti-tank role. This philosophy of increasing the weapon's flexibility is actively followed by the Russians, as shown by their 76 mm family of guns, and also the 122 mm towed howitzer.

There are two basic types of attack against armour using kinetic energy or specially adapted HE projectiles. Only guns with very high muzzle velocities can generate the energy needed for the former, but both guns and missiles can use the specialised explosives. Altogether, six types of anti-tank projectile are used by the weapons described in this book and are known as: AP, HVAP, APDS, HEAT, HESH and APHE.

AP (Armour Piercing) and HVAP (High Velocity Armour Piercing) are solid shot which penetrate armour by their kinetic energy and thrash round inside the tank.

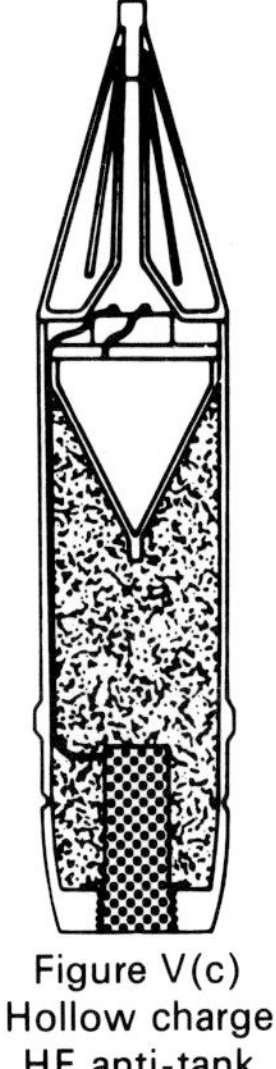

Figure V(c)
Hollow charge
HE anti-tank
(HEAT)

HE Squash
head
(HESH)

Armour piercing
HE
(APHE)

Rocket assisted
projectile
(RAP)

AP is an elderly and very limited value projectile that is no longer produced. HVAP is associated with small calibre weapons and implies the use of a special and powerful propellant charge acting on a light-weight shot with a tungsten carbide core to raise the muzzle velocity of the gun: like AP it is out-dated. An associated type of projectile is APDS (Armour Piercing Discarding Sabot) a specialist projectile fired by a few tank guns. (The only anti-tank gun to use APDS is the obsolete British 17 pounder.) The projectile is a small diameter pointed shot or dart of super-hard tungsten carbide which offers little air resistance and therefore sustains a high velocity during its flight. To fill the bore of the gun a light alloy collar, or sabot, is fitted around it and is automatically discarded on leaving the barrel. Used against light armour a solid shot may pass through a vehicle leaving it and its crew unscathed: the ideal is for the round to dissipate sufficient energy on entry to ricochet about inside.

HEAT (High Explosive Anti-Tank) consists of a shaped charge which concentrates its energy into a narrow beam and blasts a small diameter hole through armour. Its effect is to introduce a small quantity of high energy particles into the vehicle, to blind, deafen, concuss and burn the crew, and to ignite fuel and ammunition. Its efficiency depends on a good impact angle which leads to some degradation of performance as range (and fall off in the trajectory) increases. The depth of penetration is a function of the diameter of the shaped charge, hence the continued use of HVAP with small calibre weapons. The efficiency of HEAT is reduced if spin stabilisation is used. It is the usual warhead for anti-tank missiles. Some new HEAT shells fired from a rifled gun avoid spin by using a freely rotating collar in place of the usual driving band.

HESH (High Explosive Squash Head) is also known as HEP (High Explosive Plastic—not to be confused with HE/P, High Explosive/Penetrating). It is a plastic charge which

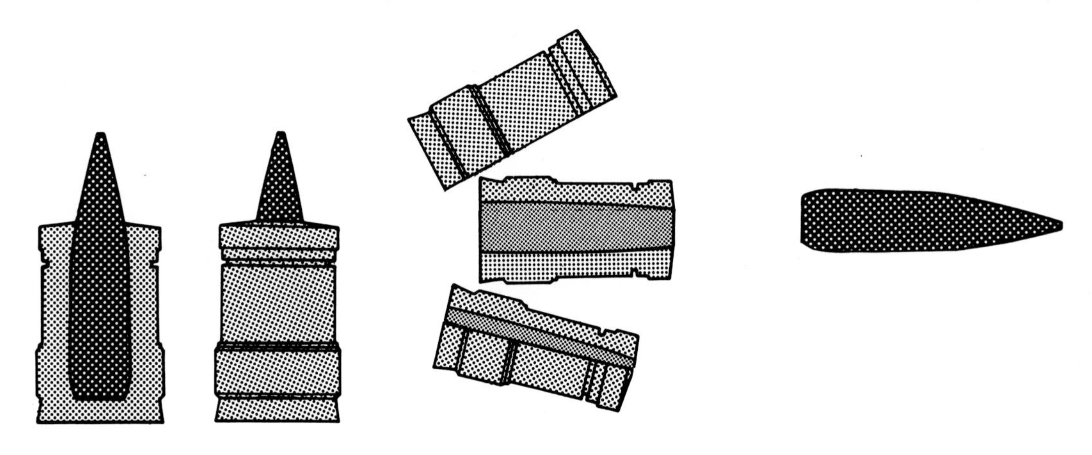

Figure V(d):
Armour piercing discarding sabot (APDS)

Figure V(e):
APDS after discharge—
shot flying free

flattens against the target before the base fuze initiates explosion. The effect is to set up shock waves in the armour which release a large scab from inside of the armour plate itself which ricochets inside the vehicle, causing crew casualities and damage to fittings. This type of shell is effective over a wide angle of impact and is therefore less subject to range limitations than HEAT. Performance is directly related to the size of the explosive and therefore the bulk of the shell. Unlike HEAT, HESH has some value as an anti-personnel weapon in an emergency as its explosive effect and the fragmentation of the shell case are lethal.

APHE (Armour Piercing High Explosive) has a similar anti-personnel effect to HESH. It is a combination of kinetic energy and explosive types of shell. Its nose is a solid cone designed to penetrate armour plate like an AP round, within which is a small high explosive charge, and behind that a base fuze. The object is to penetrate the vehicle and explode the charge inside. In practice, this is easier said than done, because of the difficulty in providing a sufficient delay in the operation fuse, and the tendency of the nose-cone to shatter, and waste the HE filling in a harmless explosion outside the tank.

Because each type of anti-armour shell has its own limitations, the various options tend to be complementary. A gun will often have more than one type of anti-armour round in its ammunition inventory.

As a result of infantry and artillery adopting armoured SP vehicles, the array of hard targets on the battlefield has increased. Faced with a predominantly armoured enemy, it is not sufficient to rely on only direct fire anti-armour weapons. Attrition must begin as far forward as possible, and the most efficient and readily available weapon delivery system is the gun. The concept of using the field gun in an indirect fire anti-armour role has produced a difficult problem. It is not sufficient to deliver an anti-armour projectile *near* a tank: a direct hit is necessary, but the zone of a field gun is too large to guarantee a hit. Furthermore, the time needed to adjust fire to the best accuracy possible against a moving target negates the whole process.

One solution lies in the development of multiple warheads which actually benefit from the inherent spread of artillery fire. The attack is against the weakest parts of an armoured vehicle—its roof or belly. A shell of suitable calibre contains either a number of 'minelets' or 'bomblets'.

A free flight rocket tends to be less accurate than a gun, but presents a formidable system for the delivery of multiple warhead weapons. It is less restricted by calibre, imposes lower firing stresses, and requires a relatively inexpensive launcher. A multiple launcher can deliver a great weight of fire very quickly, so that its delivery error is compensated for by the saturation of a large area. Speed of delivery and good delivery dispersion are important virtues against moving targets. The inter-relation of warhead and weapon design are illustrated by the design of the new West German rocket system LARS, which can deliver anti-tank minelets as well as anti-personnel warheads.

The successful use of the so called 'Smart' bomb by the Americans in Vietnam presages the development of a new long range anti-armour weapon for the artillery. Terminal guidance of a conventionally fired shell, homing on to a target illuminated by a laser beam, could soon be in service. The Americans have successfully test fired a 155 mm cannon-launched guided projectile (CLGP). At ranges of more than 10,000 m, and despite deliberate aiming errors of several hundred metres, direct hits have been achieved. The cost of each round and the associated target markers will be high, but still cost-effective against a tank especially since the delivery weapon enjoys immunity from direct retaliation.

Most shell types are designed to operate with a particular fuze, but the familiar high explosive shell can be fitted with a variety of fuzes, an examination of which illustrates the options open to both the designer and tactician. Most common is the impact or point-detonating: it provides for a combination of damage from explosion and fragments. To optimise the explosive effect on prepared defensive positions and other 'solid' targets a delay fuze can be used. It is common for one design to cater for both options by a small adjustment before loading.

HE shells can be used to cause casualties by blast, by the kinetic energy of the fragments, or by internal explosion. The type of fuze fitted determines which effect will be achieved by the shell. The cheapest and most common type is the impact fuze. Fitted with a delay device this simple fuze will allow the shell to penetrate the target before exploding: this increases the shell's effectiveness against earth-works and concrete.

Impact bursting is inefficient because a large percentage of the fragments is driven harmlessly into the ground. A better lethal effect is obtained by bursting the shell above the target. To achieve this a time-fuze which is adjustable to the time of flight is used. The older time-fuzes were of the powder-burning or clockwork types neither of which produced the accuracy or consistency needed to give the optimum height of burst. The most advanced method of obtaining an air burst is the proximity fuze which consists of a small radio transceiver set to initiate the explosion at the correct height when its signals are echoed back from the ground. The anti-aircraft 'proximity' fuze works on exactly the same principle, giving the warhead a large lethal radius and obviating the need for a direct hit.

Fuzes are technically described as 'point initiation-base detonation' (PIBD), 'point impact delay' (PID), 'proximity' (P) and 'time' (T).

Field artillery shells are normally appropriately fuzed on the gun-position to suit the fire mission, but the future concept is the PIBD fuze which will combine all these capabilities in a single complete round unchanged between factory and gun position.

The discussion of ammunition types must include the nuclear shell. The rocket has long been a prime delivery vehicle for atomic weapons. The development of a nuclear device able to withstand the shock of firing, and contained in a shell has not displaced the rocket, but complements it. The yield of gun-delivered nuclear weapons is small—a desirable quality when troop safety is critical or when the damage radius must be limited. In turn the sub-kiloton yield requires precise delivery, and the suitability of the gun with its inherent accuracy illustrates again the interdependence of warhead and weapon design.

The foregoing has dealt with the types of shell, projectile and warhead to be found in use today. The other aspect of ammunition is the propellant, i.e. the explosive substance used to propel the projectile up the bore of the gun.

A propelling explosive can be defined as a substance which, when ignited, decomposes at a controllable rate under confinement, producing high pressure gases. It is designed

Figure V(f): Types of charge.

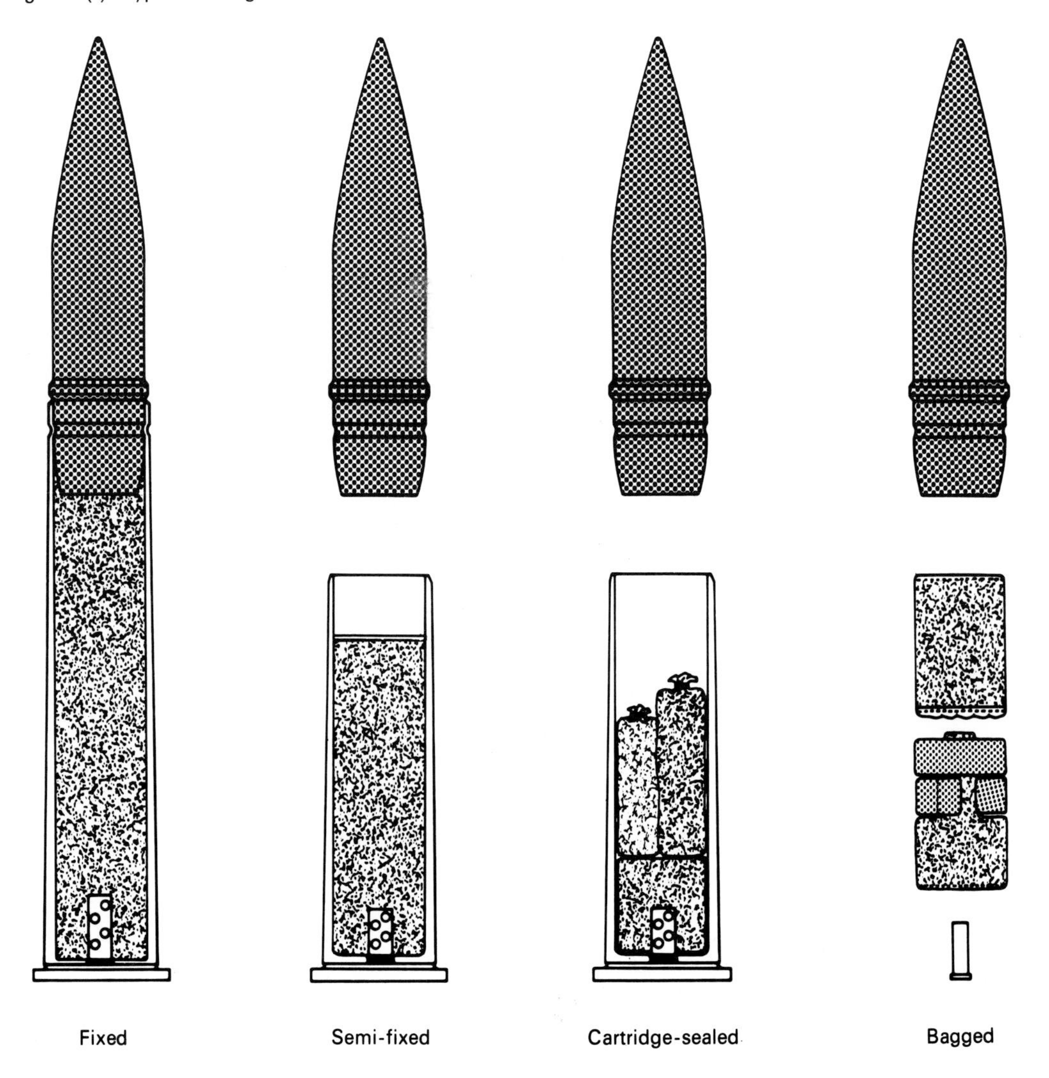

to leave no appreciable residue of solids after combustion. It should be insensitive to shock. The rate of burning should remain fairly constant under all conditions of storage and use in service and should not be affected by age, or variation in ambient temperature. When a propelling explosive in a gun chamber or rocket motor is suitably ignited by means of a primary explosive, assisted where necessary by a booster in the form of an igniter, the gas pressure will rise until the projectile commences to move. Pressures will then continue to rise until it reaches a maximum when it will decrease as the shot accelerates and so increases the volume for the chamber gases.

It is clear that the control of the burning rate affects the gas pressure, the velocity of the projectile and in turn the safety of the gun or rocket motor walls. It is possible with solid propelling explosives to control the rate of burning not only by varying the chemical composition of the propelling explosive but also by controlling the surface area. The larger the surface area the greater is the volume of gas produced in a given time. To meet the different requirements for individual weapons, propellant explosives are made in many forms such as flake, rods, pellets, balls, and small and large tubes: multi-hole tubes can be designed to increase progressively the volume of gases as the burning proceeds to correspond with the increase in area of the chamber as the shot accelerates.

The most common types of propellant can be grouped under four headings: gunpowder, single base, double and triple base and solventless (in plastic).

Liquid propellants such as alcohol/liquid oxygen or high-test peroxide/nitric-acid mixtures have been tried in guns but have not been satisfactory from the handling and safety aspects. They are used in large rockets.

Gun and ammunition designers are constantly striving to achieve increased range, so that their artillery can not only reach out to targets far behind the enemy's front lines, but also have the flexibility to fire diagonally across its own front, when required, to augment the fire support available to neighbouring formations. As much as a 30 per cent improvement in range can be achieved by two types of ammunition. First, a rocket assisted projectile (RAP) (see fig Vc) has a longer trajectory, boosted by a rocket propellant in the shell body behind the warhead. This is the most effective method in service at present, but a RAP is not so accurate, and its HE content, and hence its lethality, is less than a conventionally designed shell. Secondly, there are special streamlined extended range projectiles (ERP), which fly further because they are less affected by drag than other shells. These achieve greater accuracy than RAP, but they are less popular because they are expensive to produce and because their long, non-standard shape leads to logistic problems.

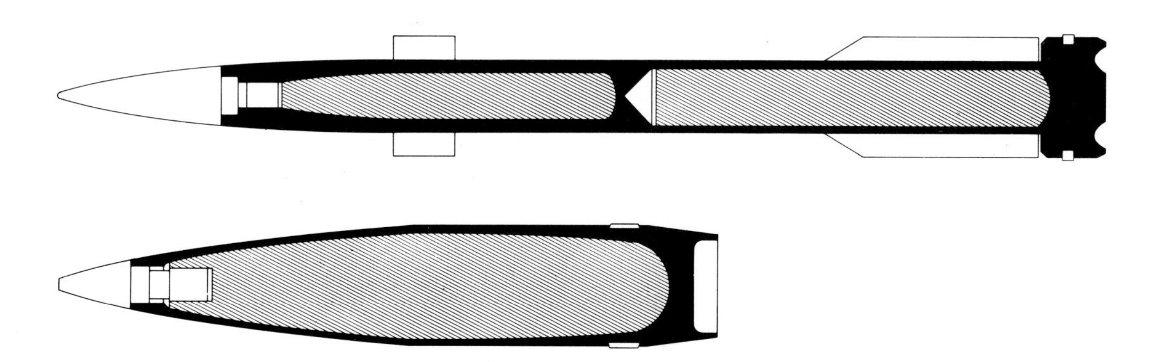

Figure V(g): Example of a long, thin, sub-calibre fin stabilised ERP projectile. Compare length and calibre with standard HE type shell for the same gun

III. Field Guns and Howitzers

The term 'field artillery' covers a group of weapons used to support ground troops. Rockets and mortars (discussed separately) fulfil a similar role and are often embraced by the term to distinguish them from air defence, coast or anti-tank artillery.

The word gun is used loosely to cover two types of equipment: the *gun* is a high velocity, flat trajectory weapon noted for its range: the *howitzer* fires a generally heavier projectile in a high and variable trajectory. In practice, the qualities of the two types are merged and some equipments—the British 25 pounder—were even called 'gun-howitzers'. In all modern equipments a set of graded propellant charges is provided with each complete round which enables the trajectory to be varied and also no more energy used, and so no more than the required shell velocity to reach the target, which reduces wear in the bore. Gun characteristics confer a secondary direct fire anti-tank capability: howitzers are easier to site to clear intervening crests (in forests or mountainous country) and can search into deep valleys.

Any heavy weapon has an anti-armour potential. This can be developed as a self defence quality, or exploited to add flexibility to the weapon's deployment. A good example is the Italian type 56 pack howitzer: an alternative mounting for the wheel axles presents either a low silhouette anti-tank gun or a field howitzer capable of high angle fire. The Russians are particularly adept at investing guns with a dual anti-tank/field artillery role. Perhaps this is in response to the fluid battle to be expected on the featureless Russian plains—in any event their artillery is deployed well forward. This and their techniques of fire control, which place the responsibility for the production of gun sight data on the observer and not the gun position command post, facilitates changes between the two roles. The result is a number of Soviet equipments which do not fall readily into a simple classification. The names by which they are known differ, and different user nations have their own deployment plans. The 85 mm M 1945 (D 44) is known as a field gun, yet its elevation is limited to 35°, its silhouette is low, and it is mounted directly on the ASU-85 as a self propelled tank destroyer. For the purposes of this section, a recognised and usual employment in the field role is the criteria for inclusion.

The essential feature of the field gun is its indirect fire capability. This, and its range potential, allows its removal from the line of contact to a position where it can exploit its all weather, 24 hour capabilities in relative immunity from direct attack. Its range can be exploited forward of the immediate battle, or laterally to assist flank units, making it a thoroughly flexible weapon system.

There are two systems for controlling indirect shooting—'observed' and 'predicted' fire. They are not mutually exclusive, because both methods are combined in modern systems so as to ensure speed and accuracy.

In the simplest form of observed fire the observer using his map gives a rough indication of the target locality and the battery immediately responds with a single round from which corrections are ordered based on observation of the fall of shot: the process is repeated until there is evidence that the mean point of impact coincides with the target and the necessary fire for effect of so many rounds from the whole battery is ordered. With good drill at the gun position and a skilled observer it is a fast and effective method, but eye-

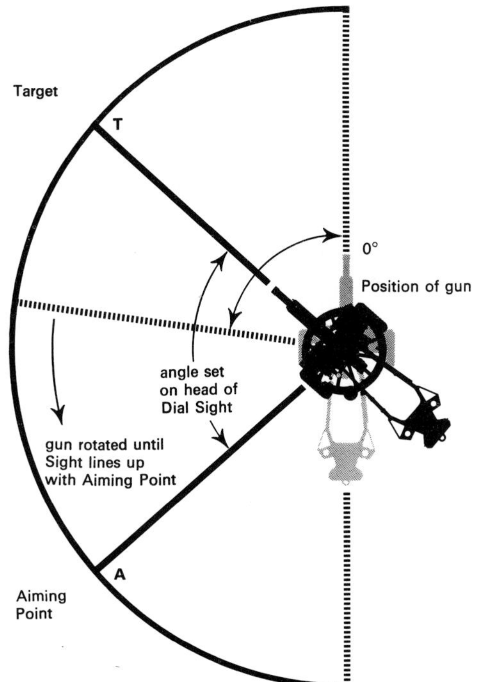

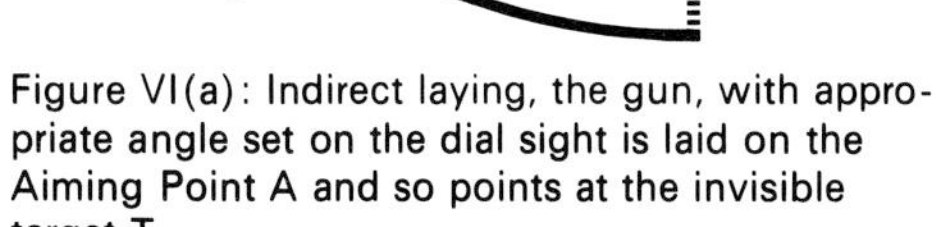
Figure VI(a): Indirect laying, the gun, with appropriate angle set on the dial sight is laid on the Aiming Point A and so points at the invisible target T

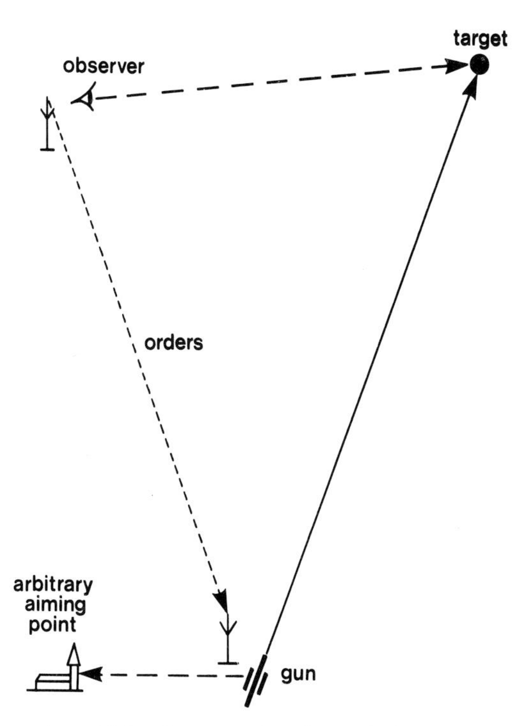

Figure VI(b): Remotely sited forward observer controls fire by radio from gun on to distant target

shooting is a knack rather than a skill and some never acquire it. It is still used for short range weapons, especially infantry mortars, and for air observation from helicopters. Its disadvantages are that the ranging, or adjusting rounds may alert the target; if, as is now often the case, the target is armoured and mobile it will be gone before adjustment is complete. At the long ranges at which modern guns are used the time of flight is long and this slows the process down.

This basic technique is retained because it is always effective in the stress of battle and proof against all system failures except the death of the observer or the failure of his communications.

The laser range finder and gyro-compass can now produce highly accurate bearing and distance information of the target for the observer, and can be used in conjunction with night observation devices. They depend for their effectiveness on the precise survey of the observer's own position. Each can be pressed into service as survey instruments for this purpose, but the practical problems of conducting a survey scheme at the forefront of the battle cannot be underestimated. One answer being developed in Great Britain is the Position and Azimuth Determining System (PADS), designed to carry survey forward using the movement of a vehicle on the inertia guidance principle which is well tried in the aerospace industry.

It has been observed from operational analysis of target engagements that target damage is nearly all inflicted in the first few seconds of fire for effect, so that the aim with 'observed-predicted' fire is to achieve three or four on-target salvos without preliminary adjustment; thus obtaining maximum effect with maximum economy of ammunition and also freeing the battery for the next mission in the shortest possible time.

Indirect fire, whether predicted or observed, requires that two pieces of information are given to the gun sight: an angle in azimuth (a bearing), and an angle of elevation. (Some equipments such as the British 25 pounder, have sights which accepted range in terms of distance which is mechanically translated into an angle of elevation: this method is not suited to the inclusion of calculated adjustments for range necessitated by predicted shooting.)

The components of the equation to produce predicted gun sight data are three-fold: those relating to the relative locations of gun and target; the gun's performance; and the variables influencing the shell in flight. All three types lend themselves to computer solution. The first group is:

1. The distance between gun and target, i.e. map range.
2. The direction of the target from the gun, i.e. map bearing.

3. The difference in elevation between gun and target, i.e. angle of sight. This difference leads to an adjustment of range.
4. The displacement of individual guns from the battery centre.
5. The rotation of the Earth beneath the shell in flight.
6. Jump, or the small displacement in the barrel's elevation as the projectile travels up the bore.

The gun performance factors are:

7. Variation in muzzle velocity from firing table data caused by wear.
8. Variation in shell weight. This can be an individual variation in manufacture, or the difference between types of projectile.
9. Charge temperature.
10. The accuracy of the gun sights.

Influences on the shell in flight vary with the time of flight, and in the case of meteorological factors, with the height of the trajectory. They are:

11. Wind.
12. Air density varying with temperature and barometric pressure.
13. The difference in the ballistic performance of different types of shell.
14. Drift, which is caused by air pressure on the rotating projectile.

The data necessary to determine the effect of these factors is available from three sources: the target and gun fixation data: calculations based on experimental firings; and continual monitoring of the meteorological environment, and gun performance. The last of these is the most difficult because of the speed with which the weather can change, and the time taken to make observations. It is not practicable to establish the atmospheric conditions throughout the shell's trajectory much of which will be over enemy territory. The sampling technique is to release a balloon equipped to measure and transmit temperature information. By observing the speed of ascent and lateral movement, the necessary data can be calculated. This time-consuming task requires highly specialised equipment, and clearly can only be done behind the fighting line. The optimum compromise is to make observations as near to the probable trajectory as possible, sighting the equipment to serve the maximum number of batteries. Up-dating must be frequent. There must be a minimum delay in the production of results, for even in the most stable climatic conditions there are times of rapid change at dawn and dusk.

One solution to the problem is the British AMETS. Its significance lies not so much in the detailed technology of the equipment and its handling, as in the computerisation of the results. These are produced as data intelligible to the gun position computer, making automatic data transfer possible.

The other variable factor which requires monitoring is the gun's muzzle velocity. In the past an assessment has been made based on the observed performance of the piece during experimental firings, or calculated from carefully conducted calibration shoots. The job is now done at the gun position, in action, electronically. The equipment works on the doppler principle. There are several systems in use.

Fourteen factors which affect the accuracy of predicted fire have been noted. The accumulation of small errors in the observation or calculation of each one can be disastrous. A one mil error in only four factors affecting bearing (gun orientation, wind speed, observed target location and data computation) can result in a 60 m delivery error at a range of 15,000 m.

It has been made apparent that the handling of the many calculations needed to predict the fall of shot can now be computerised. The increase in the speed of reaction and the high degree of accuracy brought about by computers improves the quality of artillery support, and, by reducing the time taken for each mission, increases the availability of the guns to engage more targets. An important by-project is a simplicity of operation which reduces training time and removes the fatigue of long mathematical calculations. However, command post staffs still need to understand the principles of gunnery in the event of system failure or damage in action.

The first computers handled the basic calculations. A second generation incorporated memory banks, and the latest equipment forms a complete fire control system. The American FADAC, components of which are in service, goes beyond the calculation of predicted fire data, and solves the problems of weapon and ammunition availability.

The basic premise of prediction is that the location of the target is precisely known, and that its nature is sufficiently understood to select the optimum target effect projectile. Equipment to produce this information is in service. It does not replace, but complements the observer: no one technique of target acquisition being exclusive to all others.

Of the systems which dispense with the human observer, the oldest is sound ranging. It is accurate to 100 m up to a range of one and a half times the length of the microphone base. The system has been updated by the development of radio link data transmission, but the microphones are slow to deploy. It requires more expert analysis of results than radar, but is free from enemy deception and interference except in the radio link.

Two types of radar are available: battlefield surveillance and weapon locating. Battlefield surveillance radar has been reduced in size in recent years, and gives details of any enemy movement which could lead to the presentation of a worthwhile target.

Radar can be used to tract projectiles with high trajectories and so pin-point their origin, but the problems of applying this technique to locating high-velocity, low-trajectory weapons has not so far been solved, so far as is known.

Sound ranging and radar have the benefit of being active simultaneously with the engagement of the target. A judicious adjustment to the operating procedures allows 'observation' of the fall of shot—a useful back-up facility to the prediction process.

New sources of target data are the 'drone' and the Remotely Piloted Vehicle (RPV). The existing systems, such as the Canadian USD 501, require recovery, processing and interpretation of film record after each sortie. Infra-red linescan adds to the versatility of the equipment, but the system does not produce details of target damage without flying a second sortie. In-flight film processing is a logical and time saving development, but one which may be overtaken by the introduction of television cameras. The next generation of RPVs is likely to be an automated aerial observation platform, able to hover and direct fire on the target it 'finds'.

It can be appreciated that the successful prediction of artillery fire is a highly complex matter, and subject to error at many stages. Ultimately, the result depends on the designed accuracy of the gun, and the size of its zone of dispersion. Development is now concentrating on the elimination of these errors as far as possible. In the meantime the continuing importance of the human observer cannot be over-stressed.

76·2 mm mountain gun M 48 B1

Yugoslavia

Calibre	76·2 mm
Barrel length	15·5 calibres
Muzzle brake	Multi-baffle
Trail type	Split
Weight	705 kg
Ammunition type	Semi-fixed
Charges	—
Ammunition options	HE (6·2 kg)
	HEAT
	Smoke
Rate of fire	6–7 rds/min
Muzzle velocity	398 m/sec
Range	8,600 m
Elevation limits	−15° + 45°
Traverse limits	50°
Detachment	6

This equipment is sometimes referred to as the M 48 howitzer. It is a light-weight weapon, similar in concept to the Italian pack howitzer (Model 56). Both are designed for use in difficult terrain, breakdown into pack loads, and are towed with trail legs folded. The Yugoslav gun is regularly horsedrawn.

The barrel is short, even with its multi-baffle muzzle-brake fitted, a fact which must contribute to its modest performance. The breech block is a horizontal sliding

76 mm (Yugoslav) mountain gun

wedge type. The recoil system is mounted beneath the barrel.

The trail legs are of tubular or square section and are each fitted with a vertical anchor type spade. They fold to shorten the carriage when being towed.

Several versions have been produced, with minor modifications to the basic design. One such version is used as a heavy support weapon by the infantry.

The high shield with straight top and wings is a recognition feature.

Employment	Burma
	Sri Lanka
	Yugoslavia

85 mm field gun M 52

Czechoslovakia

Calibre	85 mm
Barrel length	56 calibres
Muzzle brake	Double baffle
Trail type	Split
Weight	2,095 kg
Ammunition type	Fixed
Charges	One
Ammunition options	HE (9·5 kg)
	HVAP (5·0 kg)
	APHE (9·3 kg)
Rate of fire	15 rds/min
Muzzle velocity	805 m/sec (HE)
	820 m/sec (APHE)
	1,070 m/sec (HVAP)
Range	16,160 m
Elevation limits	−6° + 38°
Traverse limits	60°
Detachment	6

The Russian 85 mm M 1945 has been classified as an anti-tank weapon, this being its primary role. The Czech M 52 is the equivalent of that gun, and is designed to fire the same ammunition. However, it carries the clear title 'field gun', is well suited to that role and is therefore classified as such. It is manufactured at the Skoda Works in Pilsen.

The barrel is marginally longer than the Russian guns, and performance is improved accordingly. Elevation is also increased, and, although it still falls short of the optimum 45°, contributes to a better maximum range. The vertical sliding breech is retained. The recoil system, mounted below the barrel, projects forward of the shield.

The HVAP round with its larger charge, gives a muzzle velocity of 1,070 metres/second and penetrates 107 mm of armour at 1,000 m. The APHE shell penetrates 123 mm of armour at 1000 m.

The M 52 shares the distinctive wavy topped shield with its Russian equivalent. The trail is similar, with fixed spades, but a stowage box is carried across the legs when travelling.

Employment	Algeria
	Austria
	China*
	Czechoslovakia
	Egypt

* as Type 56

105 mm light gun

France

Calibre	105 mm
Barrel length	30 calibres
Muzzle brake	Double baffle
Trail type	Split
Weight	1,200 kg
Ammunition type	Separate
Charges	8
Ammunition options	HE (Mk 63) (16·16 kg) HE (M 1) (14·9 kg) HEAT (Occ-105-F1) Smoke WP Chemical Illuminating Carrier Blank
Rate of fire	—
Muzzle velocity	—
Range	15,000 m
Elevation limits	−5° + 70°
Traverse limits	45°
Detachment	—

This new gun, first seen in public in 1971, has yet to be accepted for service, but trials have reached an advanced stage. It has been designed as a high mobility dual role (anti-tank or field) weapon with the support of airborne forces in mind. It is classified with field equipment because of its similarity to the British light gun.

The long barrel with its vertical sliding breech block and high efficiency muzzle brake is mounted over the recoil system. A variable length recoil allows full use to be made of the upper register potential of the mounting: a recoil pit is not needed.

The chamber accepts American M 1 type ammunition which increases the shell option and gives logistic flexibility. The range is limited to 11,600 m using this ammunition. The gun is primarily designed to use the French Mk 63 hollow base shell and its associated charge system. The anti-armour round is the non-rotating hollow charge OCC-105-F1. It is effective to 850 m.

The trail legs are simple box section members, whose attachment to the carriage is ingenious. The legs are hinged so that the action of opening them from their travelling position lowers the carriage on to the ground. As the whole carriage rests on the ground it is completely stable. Extreme lightness confers high mobility: the gun is heli-portable, air-droppable and can be towed at speeds up to 90 km/hour. For the tow to be stable at high speed, the barrel must be traversed to the rear and clamped over the trail legs, but for short tactical moves it can travel in the firing position. It shares this dual configuration with the British light gun.

A shield has been designed, but is not always fitted.

Employment France

105 mm field howitzer (L)

Federal Republic of Germany

Calibre	105 mm
Barrel length	35·5 calibres
Muzzle brake	Single baffle
Trail type	Split
Weight	2,500 kg
Ammunition type	Semi-fixed
Charges	7
Ammunition options	HE (14·9) HEAT Smoke WP Chemical Illuminating Carrier Blank
Rate of fire	8 rds/min
Muzzle velocity	640 m/sec
Maximum range	14,500 m
Elevation limits	−5° + 65°
Traverse limits	45·5°
Detachment	8

This is a much modified American M 101 A1 produced by Rheinmetall Gmbh for the Bundeswehr.

The barrel is new and longer by 13½ calibres. A simple single baffle muzzle brake is fitted. The effect, using the same M 1 ammunition, is to increase muzzle velocity to 640 m/sec and range to 14,500 m.

The carriage and general arrangement are unchanged although weight is increased by 280 kg. There are minor modifications to the lighting for peace-time road use, which are to meet local regulations. The split trail is of conventional design. The over and under recoil system is retained.

The disadvantage arising from these modifications is that the gun is unbalanced when travelling. It can be towed for short distances only, and even so care must be taken to avoid grounding the muzzle in undulating terrain. The gun is therefore normally carried in the Faun GT 8 truck. The vehicle is modified by the addition of a simple hydraulic hoist, consisting of two booms mounted either side of the cargo space.

The weapon is used in the general artillery reserve of a corps, but is soon to be phased out of service.

Employment Germany (FRG)

(illustrated overleaf)

105 mm field howitzer (L). The German adaptation of the American M 101 A1

105 mm field howitzer

India

A new 105 mm field howitzer is being produced by an Indian ordnance factory. It is described as an interim design and little technical data is available.

It has a light box section split trail. A massive muzzle brake is fitted, and the variable length recoil system is mounted below the barrel. A steeply raked shield is fitted, with provision for direct fire sights. The carriage is remarkably similar to that of the Italian 105 mm pack howitzer.

This picture shows the massive muzzle brake. The recoiling parts seem new

Pack howitzer 105/14 model 56

Italy

Calibre	105 mm
Barrel length	14 calibres
Muzzle brake	Multi-baffle
Trail type	Split
Weight	1,273 kg
Ammunition type	Semi-fixed
Charges	7
Ammunition options	HE (14·9 kg) HEAT Smoke WP Chemical Illuminating
Rate of fire	8 rds/min
Muzzle velocity	420 m/sec
Maximum range	10,575 m
Elevation limits	−7° + 65°
Traverse limits	18°–28° depending upon elevation
Detachment	6

This outstandingly successful howitzer has been exported to 23 countries since coming into service in 1957. Its chief virtues are its versatility, lightness and cost effectiveness. The design is a short barrel, split trail howitzer which breaks down to 12 pack loads.

The need to minimize the weight of each component part for pack transport demanded a short barrel of 14 calibres; even so it weighs 122 kg. Although the gun was designed to fire the universally available American M 1 105 mm howitzer ammunition, the length of barrel is insufficient to make full use of its potential range.

The box section trail legs are each made in three sections which are folded for travelling. The wheel stations have high and low mountings for normal firing or a low silhouette anti-tank configuration. The weapon fires in both high and low angle without preparation. A second elevating hand wheel provides a split function laying facility, normally used for anti-tank shooting.

The pack loads are animal- or man- and air-portable. All these methods (except pack animal) tend to be uneconomical in effort, but provide almost limitless flexibility in deployment over every sort of terrain. Dismantling and reassembling the weapon is quick and simple: the only tool required is a lifting yoke which clamps to the shields for travelling. More usual is deployment of the assembled gun. It flies well beneath a Wessex class helicopter, is air-droppable, air-portable and can be towed conventionally by a Land-Rover type vehicle. For long road journeys it is normally carried portee in a 3-ton truck. Skis can be fitted to the wheels for towing over snow.

A special German version features a redesigned single-baffle muzzle brake to replace the multi-baffle type, and the Spanish use a version which looks similar but is reputedly built under licence in Spain.

Employment

Argentina	France	Pakistan
Australia	Germany (FRG)	Saudi-Arabia
Austria	Ghana	Spain
Bangladesh	India	Sri Lanka
Belgium	Italy	Sudan
Canada	Malaysia	Zambia
Chile	Nigeria	Zimbabwe
Eire	New Zealand	

The howitzer, in action after helicopter delivery with the lifting harness still fitted. Note the yoke used in dismantling and reassembling fixed to the shield

The same—at maximum elevation for high-angle fire

The German version with muzzle brake in anti-tank mode

Light howitzer 105/26

Spain

Calibre	105 mm
Barrel length	26 calibres
Muzzle brake	Double baffle
Trail type	Split
Weight	1,950 kg
Ammunition type	Semi-fixed
Charges	—
Ammunition options	HE (15·27 kg) HE (14·9 kg) HEAT Smoke WP Chemical Illuminating
Rate of fire	4 rds/min (see text)
Muzzle velocity	443 m/sec
Maximum range	9,400 m
Elevation limits	−5° + 45°
Traverse limits	50°
Detachment	6

The 105/26 is a Spanish weapon of post-war design (1950) which fires American M 1 ammunition as well as its own type of shells. The American round weighs 0·3 kg less, has a lower muzzle velocity and therefore produces less range.

Note the simplicity of the trails with folding traversing levers

The monoblock barrel is fitted with a double baffle muzzle brake. The breech has a horizontal sliding block. The hydropneumatic recoil system is mounted below the barrel. Sighting is conventional, to the left of the barrel; a second elevating hand wheel is fitted to allow split function laying.

The carriage is a conventional split box section trail running on large section tyres. A hand spike is built in to each trail leg.

The sustained rate of fire is quoted as 4 rounds/minute, but this can be doubled for short periods to give a burst of fire effect.

Employment Spain

---

105 mm light field howitzer 4140

Sweden

Calibre	105 mm
Barrel length	28 calibres
Muzzle brake	Single baffle
Trail type	Cruciform
Weight	2,600 kg
Ammunition type	Semi-fixed
Charges	8
Ammunition options	HE (15·5 kg)
Rate of fire	8 rds/min (see text)
Muzzle velocity	610 m/sec
Maximum range	14,600 m
Elevation limits	−5° + 65°
Traverse limits	360°
Detachment	6

Something of the Bofors' expertise with the design of air defence guns has been incorporated into the design of this gun. The carriage is designed for all round traverse, and features four legs in a 'quadruple span'.

The barrel tapers to a stubby single baffle muzzle brake. The recoil system, mounted below it extends nearly to the muzzle. The weight of the recoil system being well forward and the rearward position of the trunnions makes it necessary to fit equilibrators which run forward

The howitzer in action

from a rear mounting over the saddle.

A very high burst-rate of fire of 25 rounds/minute is possible, partly because semi-fixed ammunition needs only a single action to load, and partly because of the excellent access to the breech. The sustained rate of fire is eight rounds/minute.

In action, the road wheels remain fixed to the carriage, but are raised to clear the trail legs: the arrangement is similar to the Russian 122 mm D 30. High angle fire up to 65° is possible; the height of the trunnions, which are well to the rear of the barrel, prevents any loss of elevation when the barrel is in line with a leg. Each leg is flattened into a small pad at its end, and accepts a vertical ground spike, which acts as a spade. Vertical firing loads are passed to a firing jack beneath the centre of the cruciform. As might be expected, this arrangement makes it a heavy gun for its calibre. A shaped shield is fitted behind the wheels.

The Swiss refer to this weapon as the 105 mm gun 35; it is also known as the Bofors L/28.

Employment Sweden
Switzerland

---

105 mm gun M 35

Switzerland

Calibre	105 mm
Barrel length	42 calibres
Muzzle brake	Double or treble baffle
Trail type	Split
Weight	3,840 kg
Ammunition type	—
Charges	—
Ammunition options	HE (15·3 kg)
Rate of fire	5 rds/min
Muzzle velocity	800 m/sec
Maximum range	21,000 m
Elevation limits	−5° +45°
Traverse limits	60°
Detachment	6

Like the M 46 howitzer, the M 35 is a Bofors design made under licence in Switzerland. It is very elderly and may still have the original metal spoke, solid tyre wheels.

The long tapered barrel is mounted over the recoil system. There is a square, upright shield which overlaps the wheels. A semi-cylindrical counter weight merges into the breech ring, which is fitted with a horizontal sliding block. The straight trail legs end in angled hand spikes with spades fitted beneath them.

Performance has been improved by the introduction of a new ammunition. Muzzle velocity is raised to 800 metres/second from 785 metres/second by a bigger charge, and range is improved from 17,500 m to 21,000 m. Stocks of the older ammunition are still being used.

Employment Switzerland

The 105 mm M35 gun in action

The M 35 in the travelling position—fitted here with solid tyred wheels

105 mm field howitzer M 46

Switzerland

Calibre	105 mm
Barrel length	22 calibres
Muzzle brake	Multi-baffle
Trail type	Split
Weight	1,850 kg
Ammuntion type	—
Charges	—
Ammunition options	HE (15·15 kg)
Rate of fire	—
Muzzle velocity	490 m/sec
Maximum range	10,000 m
Elevation limits	0° + 65°
Traverse limits	60°
Detachment	6

This gun is of Swedish Bofors design, made under licence in Switzerland up to 1953. It is distinguished by its five baffle muzzle brake and the recoil gear mounted under the barrel—the cradle containing it reaches nearly to the muzzle. The straight box section legs are fitted with seats for the layer and the loader. Laying arrangements are to the left of the barrel, behind the curved-top shield.

It was replaced in mechanized divisions by the M 109U, but remains in service with infantry formations.

105 mm howitzer M 46

Employment Switzerland

105 mm light gun

United Kingdom

Calibre	105 mm
Barrel length	30 calibres
Muzzle brake	Double baffle
Trail type	Box
Weight	1,818 kg
Ammunition type	Separate
Charges	7
Ammunition options	HE (16·05 kg) Smoke BE and WP (15 kg) Coloured smoke (15 kg) Illuminating (15 kg) HESH
Rate of fire	6 rds/min
Muzzle velocity	708 m/sec
Maximum range	17,000 m
Elevation limits	−5° + 70°
Traverse limits	360° (top traverse 10°)
Detachment	6

The 105 mm light gun is designed to replace the Italian pack howitzer Model 56 with a modern more robust but light air-portable weapon. The requirement for improved range has been met by using Abbot Mk II ammunition. The weapon has been accepted for service, and is now in full production.

The autofrettaged barrel is 30 calibres in length. The vertical sliding block is fitted with an electric firing mechanism. A special barrel which fires American M 1 ammunition is available; it can be installed in two hours, but having a limited range performance will be used for training only. The trunnions are set well to the rear of

Note the bowed trails and detail of sights and breech

The 105 mm light gun in its battlefield towing configuration, with the barrel in the firing position. For long-distance travel the piece is rotated and clamped to the trail. The towing vehicle is a 1 tonne Landrover 4 × 4

105 mm light gun in action

the barrel. Although the centre of gravity of the breech and recoil system is close to them equilibrators are necessary. They are simple exposed coil spring types, and are mounted almost horizontally beside the recoil system. The recoil system is hydro-pneumatic and incorporates a cut-off gear which halves recoil length at full elevation. (1·14 m at 0° elevation, 0·5 m at 70° elevation.) It is mounted under and over the barrel.

The Abbot Mk II ammunition has a 25 per cent more lethality than the American M 1 type. The HE round can be fitted with the L32 percussion fuse, the L27 proximity fuse or the L33 fuse which combines mechanical time and impact detonation. The carrier-type shells are ballistically matched to the HE round. While the special barrel is needed to use M 1 ammunition complete, the M 1 shell only can be fired with a special charge system in the normal barrel.

Both the direct and indirect fire sights are illuminated by 'Trilux' nuclear light sources and the laying system is fully compatible with present and projected British fire control systems.

The box trail of hollow tubular members, and circular platforms are reminiscent of the 25 pounder. In plan view, the trail members have a curved shape. The arrangement of the carriage saves 150 kilogrammes over a split trail design, and allows speedy 360° traverse. Important features of the design are the advanced material used (a new high quality steel), and the construction methods; welds have been placed only in low stress areas and explosive forming is used to shape complex sheet steel members.

Stability is provided by the platform, but a four spade system is available, viz:

1. A large flat sole plate, used with the platform, to allow easy traverse of the trail.
2. A field spade (the sole plate rotated) for use without the platform.
3. A shallow, rock spade.
4. Digger attachment to the rock spade to improve stability in soft ground.

The gun is normally towed with the barrel rotated to the rear to clamp over the towing eye. This compact folded configuration adds to towing stability and is used for air movement, and for air dropping on a medium stress platform. Rotation requires the removal of a road wheel and adds one minute to the into/out-of action time. For battlefield movement an unfolded towing configuration is possible. The complete gun can be lifted by Puma class helicopters: to fly beneath the Wessex, the carriage and piece are separated and reassembled in the field in ten minutes, with the help of one simple tool.

A trailing arm suspension with laminated torsion bars and shock absorbers is used to reduce the stress of cross-country movement. It stays in operation during firing and is used to assist stability. Brakes are of the hydraulic over-run type with a hand brake which can operate on each wheel independently. Light-weight wooden or alluminium skis are under development for over snow travel.

The British light gun is heavier than some of its rivals, but is now a proven and robust design, offering a remarkable performance for its size and weight. The ratio muzzle energy/weight is 640, compared to the Russian D 30 (537), American experimental XM 164 (370) and 25 pounder (280).

Employment Entering service in UK

105 mm howitzer M 101 A1

United States of America

Calibre	105 mm
Barrel length	22·5 calibres
Muzzle brake	Nil
Trail type	Split
Weight	2,220 kg
Ammunition type	Semi-fixed
Charges	7
Ammunition options	HE (14·9 kg) HEAT HE/P Smoke WP Chemical Illuminating
Rate of fire	3 rds/min
Muzzle velocity	473 m/sec
Maximum range	11,160 m
Elevation limits	−5° + 66°
Traverse limits	45°
Detachment	8

The M 101 A1 is one of the most successful guns ever produced. Between 1940 and 1953, 10,202 were built and supplied to 46 armies. Development of a weapon known as the M 1 started as early as 1928, but by 1940 only 14 had been made. It is a contemporary of, and rival to, the British 25 pounder, which is lighter by 500 kg, outranges it by 2,000 m, but fires a lighter shell of only 11·34 kg.

The barrel is short, a fact which contributes to the modest range. It has a life of 20,000 EFCs. A horizontal sliding breech block is used with percussion firing mechanism. The hydro-pneumatic recoil system, mounted over and under the barrel, is a constant recoil length type.

The M 1 ammunition system was designed for this gun and has become a standard NATO ammunition usable by a number of more recent equipments, including the Abbot, French light gun, and Italian pack howitzer, model 56. Several HE shells are available (M 1, M 3, M 413 and M 444). The HE/P round penetrates 102 mm of armour at a range of 1,500 m. Among experimental shells are a CS gas round (XM 629): Beehive (XM 546), and a new HEAT projectile (XM 622). Two rocket assisted projectiles have been designed (XM 482, XM 548). Various fuses, time, proximity and percussion, were developed. The semi-fixed ammunition design speeds loading and leads to a high rate of sustained fire. The quoted rate can be increased for short periods.

The layout of the carriage is conventional and simple. The box-section, split-trail legs are straight and carry recoil stresses to built-in spades. The gun fires off its road wheels, above which the elevating mass is trunnioned well to its rear; a spring equilibrator is mounted beneath the rear cradle extension.

A number of experimental and modified weapons have been produced from the basic design. These are listed below. The M 101 A1 has now been replaced in the United States' Army by the M 102.

M3: A war-time expedient for airborne use. Barrel length was reduced to 15½ calibres and the 75 mm field howitzer carriage used. Weight was reduced to 1,134 kg

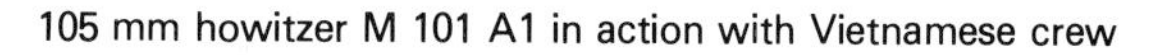
105 mm howitzer M 101 A1 in action with Vietnamese crew

105 mm HM 2. The French adaptation of the American M 101 A1. Note the new muzzle brake

The 105 mm M 101 A1 howitzer in action on a hill-top fort in Vietnam—barrel at maximum depression

and range cut to 7,580 m since the full seven charges could not be used.

M7 (Priest): An SP mounting developed during World War II on the M 3 Grant chassis. It is no longer in service.

XM 124: An SP version featuring 'Major/Minor' wheels: three balloon-tyred wheels are mounted radially around the axle. On level surfaces two wheels in contact with the ground rotate about their own axles; to cross obstacles the whole wheel assembly rotates in a 'walking' motion. Power is provided by a pack mounted, with a driver, on the trail. The trail and motor unit are supported on a dolly wheel. Lockheed Aircraft Service Company carried out the work on an M 101 A1 less shield. Performance of the gun is unchanged.

HM 2: The French used the M 101 A1 fitted with the barrel of the AMX 105/50 SP howitzer. Later, the weapon was modernised by the fitting of the 30 calibre barrel used in the AMX 105-B SP howitzer. Performance for both versions is the same as the SP weapons. A further adaptation of the HM 2 was produced during the development of the French light gun.

105 mm C 1: This is the Canadian version fitted with a new barrel and breech ring. The barrel is of auto-frettaged monobloc construction and made by Sorel Industries of Quebec. Deliveries started in 1955.

105 mm field howitzer (L): This German modification is described in a separate entry.

Employment

Australia	Greece	Morocco
Austria	Guatemala	Netherlands
Argentina	Haiti	Norway
Bangladesh	Italy	Pakistan
Belgium	Indonesia	Pathet Lao
Bolivia	Iran	Peru
Brazil	Israel	Phillipines
Canada	Japan	Portugal
Chile	Jordan	Spain
Columbia	Kampuchea	Taiwan
Denmark	South Korea	Thailand
Ethiopia	Laos	Turkey
Dominican Republic	Liberia	USA
France	Libya	Venezuela
Germany (FRG)	Mexico	Vietnam

105 mm howitzer M 102

United States of America

Calibre	105 mm
Barrel length	333 cms
Muzzle brake	Nil
Trail type	Box
Weight	1,496 kg
Ammunition type	Semi-fixed
Charges	7
Ammunition options	HE M 1 (14·9 kg) HE XM 482 (12·7 kg) HEAT HE/P Smoke WP Chemical Illuminating Flechettes (Beehive)
Rate of fire	3 rds/min (see text)
Muzzle velocity	494 m/sec
Maximum range	11,500 m
Elevation limits	−5° + 75°
Traverse limits	360°
Detachment	8

This weapon was conceived to replace the highly successful M 101, and primarily designed for air-transportability. The result is a weapon which is air-portable and airdroppable and weighs only 1,496 kg: this is at least 1,000 kg lighter than any previous American guns of the same calibre. Since its introduction to United States Airborne formations in 1965 both French and British light guns have been announced. The former is even lighter, the other outranges it by 3,500 m but at a cost of 300 kg. Many other 105 mm guns and howitzers (including the Italian Model 56) appear to rival the lightness of these new generation weapons, but none approach their range.

The barrel is 12 per cent longer than that of the M 101 A1, which increases range by 500 m using the same ammunition. A muzzle brake was fitted to early development guns, but it is not used for the production model. A vertical sliding wedge breech is fitted. The barrel is mounted at its extreme rear and two exposed spring equilibrators are mounted on the base of the saddle.

The hydro-pneumatic recoil system is fitted over and under the barrel. To make best use of the high-angle fire potential, avoiding the need for a recoil pit, the maximum recoil of 1,275 mm is reduced by half at full elevation.

Two ammunition types are used: the M 1 system and a new XM 482 long-range (rocket assisted) round which will increase the maximum range to 15,000 m. In addition to the wide variety of M 1 shells, a flechette projectile has been developed. Burst fire of 30 rounds in three minutes is possible.

The M 102 has a deep section aluminium box trail. A circular base plate travels beneath the trail and is lowered to raise the wheels clear of the ground for firing. At the rear of the trail a longitudinal roller fitted with a 'Terra-Tire' replaces the usual spade. The roller allows rapid one-man traverse through 360°. The weapon is only 1·37 m high in action at zero elevation.

Employment Kampuchea
Saudi-Arabia
USA
Vietnam

The firing platform cannot be seen in this picture, but the raised road wheel is clear, as is the deep section of the trail members

105 mm howitzer M 56

Yugoslavia

Calibre	105 mm
Barrel length	27·9 calibres
Muzzle brake	Multi-baffle
Trail type	Split
Weight	2,180 kg
Ammunition type	Semi-fixed
Charges	—
Ammunition options	HE (15 kg) HEAT Smoke Illuminating
Rate of fire	6–7 rds/min
Muzzle velocity	570 m/sec
Maximum range	13,000 m
Elevation limits	−12° + 68°
Traverse limits	52°
Detachment	6

This weapon consists of a modified American M 101 A1 barrel made under licence and mounted on a locally designed carriage. The horizontal sliding breech is retained, but range performance is considerably improved to a claimed 13,000 m. Since no other weapon using M 1 ammunition achieves such a range, it can be assumed that a new ammunition type has been developed. The over and under recoil system is retained, but otherwise the appearance is quite unlike the original equipment. A multi-baffle muzzle brake is fitted together with a shaped and sloping shield. The spades are mounted to the trail legs at their centre—not below the legs as is more common. The M 56 is just lighter than the M 101 A1.

Employment Yugoslavia

122 mm Tampella field gun M 60

Finland

Calibre	122 mm
Barrel length	50 calibres
Muzzle brake	Single baffle
Trail type	Split
Weight	9,500 kg
Ammunition type	—
Charges	—
Ammunition options	HE (25 kg)
Rates of fire	—
Muzzle velocity	950 m/sec
Maximum range	25,000 m
Elevation limits	−5° + 50°
Traverse limits	90°
Detachment	8

This is a very heavy equipment for its calibre, but it does have an excellent range. Principal employment is with armoured brigades, but it is also deployed in the coast defence role. Its general layout is similar to the French 155 mm M 50 howitzer.

The barrel is long, probably in the order of 50 calibres. It has a horizontal sliding breech block and a simple muzzle brake. The trunnions are at the rear of the barrel, which is balanced by equilibrators that are mounted onto a vertically extended saddle. The buffer recuperator is fitted over and under the barrel.

Split function laying is used.

The most unusual feature of this gun is the arrangement of the carriage. The split trail is of very sturdy construction with massive box section legs. At the rear of each leg is a pad, slotted to receive three anchor pins. These are hammered into the ground vertically—a time consuming task which must be very difficult in rocky or frozen terrain. Vertical firing stresses are passed into a firing jack beneath the cradle: The tandem road wheels are lifted clear of the ground in action. The wheels run on axles fixed directly to the trail legs: thus they only come parallel when the legs are clamped together. The legs are hinged well forward, which allows them to open wide enough to permit a top traverse of 90°. Reputedly 360° traverse is possible with 'additional equipment', although the nature of this equipment is not specified. The barrel rotates to clamp over the trail for travelling.

Employment Finland

122 mm field gun 122/46

Spain

Calibre	122 mm
Barrel length	—
Muzzle brake	Double baffle
Trail type	Split
Weight	7,800 kg
Ammunition type	—

Charges	—
Ammunition options	HE (22·1 kg)
Rate of fire	—
Muzzle velocity	830 m/sec
Maximum range	19,900 m
Elevation limits	−4° + 65°
Traverse limits	57°
Detachment	8

Nothing is known about this equipment except the data in the table above. It does not seem to be related to any other weapon of this calibre, but the choice of calibre may well have been influenced by the availability of ammunition.

Employment Spain (Reserve Army only)

122 mm field howitzer M 1938 (M 30)

Union of Soviet Socialist Republics

Calibre	122 mm
Barrel length	22·7 calibres
Muzzle brake	Nil
Trail type	Split
Weight	2,500 kg
Ammunition type	separate
Charges	9
Ammunition options	HE (21·8 kg) HEAT (14·1 kg) Smoke (22·4 kg) Illuminating (21 kg) Chemical
Rate of fire	5–6 rds/min
Muzzle velocity	515 m/sec
Maximum range	11,800 m
Elevation limits	−3° + 63½°
Traverse limits	49°
Detachment	8

This equipment dates from World War II and is still in service in the Soviet Army, despite the appearance of the new D 30 and D 74 122 mm weapons. It was built in great numbers and widely exported. It was the standard organic 122 mm weapon of the Russian Divisional and Army Group Artillery.

The barrel is not fitted with a muzzle brake, and the breech mechanism is based on the Schneider pattern of screw block, though the ammunition uses a metal cartridge case. Firing is by a self-cocking mechanical lock actuated by a lanyard. The recoil system is mounted above and below the barrel, a system copied from German practice

The HEAT round penetrates 200 mm of armour, but anti-tank range is limited to 630 m. This round weighs only 14·1 kg against 21·8 kg for the HE shell.

The split trail is also used for the 152 mm M 43. The box section legs are of rivetted construction. The spades fold for travelling as do the permanently fitted hand spikes. The gun can be fired with the trail legs closed in their travelling position, but only 1½° of traverse is possible. The large disc wheels with narrow tyres are replaced by smaller wheels in the East German Army, and by spoked wheels in Bulgaria. The steeply arched shield has a sliding centre section.

There is a Chinese built version—'Type 54'.

122mm field howitzers M 1938 (M 30) in action. Note the Soviet practice of rigid linear deployment

Employment

Albania	Finland	North Korea
Algeria	Germany (GDR)	Poland
Bulgaria	Hungary	Rumania
China	Iraq	USSR
Cuba	Lebanon	Vietnam
Czechoslovakia	Libya	Yugoslavia
Egypt		

122 mm field gun M 1955 (D 74)

Union of Soviet Socialist Republics

Calibre	122 mm
Barrel length	45 calibres
Muzzle brake	Double baffle
Trail type	Split
Weight	6,000 kg
Ammunition type	Separate
Charges	—
Ammunition options	HE (25·5 kg) APHE (25 kg)
Rate of fire	6–7 rds/min
Muzzle velocity	800 m/sec (HE) 950 m/sec (APHE)
Maximum range	21,000 m
Elevation limits	−2° + 50°
Traverse limits	60°
Detachment	9

The D 74 replaced the 122 mm A 19 in the Soviet Army, but is in turn older than the D 30. It is used for the general support of infantry and in mechanised divisions.

The long barrel is mounted low on its carriage. The vertical sliding wedge type breech is semi-automatic. The buffer/recuperator is mounted above the barrel, well to the rear so that it hardly projects beyond the shield: the effect must be to balance the barrel, which is long and mounted well to the rear.

The variable charge is fired from a cartridge case which is separately loaded. The shell seems to be that used for the A 19: the usual shell options for Soviet 122 mm equipments are available. The APHE round is effective to 1,200 m: it weighs less than the HE round and gives a muzzle velocity of 950 m/sec. It penetrates 130 mm of armour at a range of 1,000 m.

The split trail with fixed spades is of conventional design, and is very similar to that of the 152 mm D 20. Each trail leg is fitted with a castor wheel to assist in deployment and man handling. Built-in handles run forward from these wheel mounts. In action a firing jack lets down between the wheels forward of the shield and absorbs vertical firing stresses. It is a heavy gun, and the handling provisions and 9 man detachment suggest it may be difficult to deploy and operate.

The wavy-topped shield is a recognition feature. The low silhouette and effective APHE round give the weapon a useful secondary role as an anti-tank gun.

Several tractors are used, notably the AT-S, AT-L and URAL 375 (6 × 6).

Employment

Bulgaria	Germany (GDR)	Rumania
China	Hungary	USSR
Cuba	Nigeria	Vietnam
Egypt	Poland	

122 mm field gun M 1955 (D 74)

122 mm howitzer D 30

Union of Soviet Socialist Republics

Calibre	122 mm
Barrel length	33 calibres
Muzzle brake	Multi-baffle
Trail type	Tri-legged
Weight	5,000 kg
Ammunition type	Separate
Charges	2
Ammunition options	HE (21·3 kg)
	HEAT (14·1 kg)
	Smoke
	Illuminating
Rate of fire	7–8 rds/min
Muzzle velocity	690 m/sec (HE)
	740 m/sec (HEAT)
Maximum range	15,300 m
Elevation limits	−5° + 65°
Traverse limits	360°
Detachment	8

The D 30 first entered service in 1967.

The barrel is longer than usual in a howitzer and this, together with the use of a highly efficient muzzle brake, allows a powerful charge to be used to achieve a reasonable range. The breech mechanism uses a sliding semi-automatic block, while the recoil system is distributed round the ring cradle and protected by a steel shield above the gun.

Only two charges are used: they are contained in large cartridge cases, and their size is reflected in the high muzzle velocity. While the shell weights and options suggest that they derive from the M 30 the HEAT round is certainly new. It is non-rotating, fin stabilized and penetrates 460 mm of armour at 1,000 m. Muzzle velocity for this round is 740 m/sec. The gun was used extensively in the anti-tank role during the Yom Kippur War—and to good effect.

The most interesting feature of the design is its three legged trail which gives 360° traverse. For travelling, the trail legs fold beneath the barrel (by which the gun is towed). For firing, the legs deploy at 120° intervals and the wheels are jacked up high enough to clear the legs when the barrel rotates about its centre pivot. It is doubtful if full elevation is possible when the breech is immediately over a trail-leg. Pads at the end of each leg accept perforated anchor pins which act as spades. A tiny shield is fitted between the wheels.

Employment

Cuba	Germany (GDR)	Poland
Czechoslovakia	Hungary	Syria
Egypt	Nigeria	USSR
Finland		

122 mm howitzers D 30. Note Y-form train in action permitting all-round traverse. The projection below the muzzle brake is the towing eye—the gun being towed by the piece

130 mm field gun M 1946

Union of Soviet Socialist Republics

Calibre	130 mm
Barrel length	50 calibres
Muzzle brake	Pepper-pot
Trail type	Split
Weight	8,618 kg
Ammunition type	Separate
Charges	—
Ammunition options	HE (33·4 kg) APHE (33·4 kg)
Rate of fire	6–7 rds/min
Muzzle velocity	930 m/sec
Maximum range	27,000 m
Elevation limits	$-2\frac{1}{2}°$ $+46°$
Traverse limits	50°
Detachment	9

The M 1946 is a versatile equipment with a naval background and formidable anti-tank capability. It was developed from the M 1936 naval gun and replaced the 122 mm A 19 in the Soviet Army.

The barrel is over 7 m long (7·6 m including the pepper-pot muzzle brake). Muzzle velocity is 930 m/sec and results in the remarkable range of 27,000 m. Great claims are made for the accuracy of the gun at long ranges. The breech is of the horizontal sliding wedge-type.

The recoil system is mounted over and under the barrel. There is a distinctive collar support for the fixed end of the recuperator at the front of the cradle, well in front of the shield.

Ammunition is of the separate type, the variable charge being fitted in a conventional cartridge case. The APHE round penetrates 170 mm of armour at 1,000 m.

The sturdy deep box section split trail is mounted on two road wheels. The removable spades travel on top of the trail legs. A limber with two small disc wheels supports the trail for travelling. The barrel is retracted for travelling, and the breech sits between the trail legs not over them. A chain and pulley system is mounted on the trail, to retract the barrel. The small shield is not always fitted.

The range and shell size make this an ideal counter-battery weapon. During the Yom Kippur War it was used extensively in this role by the Egyptian Army. An automated command post with computer fire control equipment was linked to a ground radar system. This system, developed by the Russians, uses SNAR-2 X-Band fire control radar which has a range of 9,000 m. Small Yawn and Porktrough radars are also used.

130 mm field gun M 1946

The gun is built 'under licence' in China and it is this version which has been supplied to Pakistan, and India. There is also a later, 1954 model with a longer barrell (55 calibres); improved construction has made this gun lighter (7,500 kg), but its ballistic performance is the same.

Employment Bulgaria, China, Cuba, Egypt, Finland, Germany (GDR), India, Iran, Iraq, Israel, North Korea, Mongolia, Nigeria, Pakistan, Poland, Syria, USSR, Vietnam, Yugoslavia

Same gun, note split trail

152 mm howitzer M 1943 (D 1)

Union of Soviet Socialist Republics

Calibre	152 mm
Barrel length	25 calibres
Muzzle brake	Double baffle
Trail type	Split
Weight	3,600 kg
Ammunition type	Separate
Charges	—
Ammunition options	HE (39·9 kg) Semi-AP (51·1 kg)
Rate of fire	3–4 rds/min
Muzzle velocity	508 m/sec (HE) 432 m/sec (Semi-AP)
Maximum range	12,400 m
Elevation limits	−3° + 63·5°
Traverse limits	35°
Detachment	7

This elderly weapon shares a carriage with the 122 mm M 30 howitzer.

The barrel, fitted with a muzzle brake, is from the M 10, and the screw type breech in its square section breech ring is retained: it is cartridge obturated. The hydraulic buffer and hydro-pneumatic recuperator are mounted under the barrel.

The semi-AP round is more than eleven kilogrammes heavier than the HE shell, and has a much lower muzzle velocity: armour penetration is 82 mm at 1,000 m.

152 mm field howitzers M 1943 (D 1) in action

The slim box section trail legs have folding spades and lifting handles. It is not clear if this weapon can be fired without splitting the trail like the M 30 (which uses the same carriage). The road wheels are large diameter metal discs. The shield is steeply raked to the rear. Like other Soviet howitzers the D 1 fires at high angle without preparation.

The equipment is noticeably lighter than other guns of similar calibre—a fact reflected in its modest range. It has been phased out of service with the Soviet Army.

Employment Albania, China, Egypt, Germany (GDR), Hungary, Poland, Syria, Vietnam

152 mm howitzer M 1955 (D 20)

Union of Soviet Socialist Republics

Calibre	152 mm
Barrel length	37 calibres
Muzzle brake	Double baffle
Trail type	Split
Weight	5,900 kg
Ammunition type	Separate
Charges	—
Ammunition options	HE (48 kg) APHE (48·8 kg) Chemical
Rate of fire	4 rds/min
Muzzle velocity	655 m/sec (HE) 600 m/sec (APHE)
Maximum range	18,000 m
Elevation limits	−2° + 63°
Traverse limits	60°
Detachment	8

The D 20 152 mm howitzer is mounted on the same carriage as the D 74 122 mm gun. The D 20 barrel is shorter and thicker than the D 74 but is fitted with a similar muzzle brake. The breech block is of the semi-automatic vertical sliding wedge type. The breech is cartridge obturated. The recoil system is mounted above the barrel and projects beyond the shield.

The APHE round is 0·8 kg heavier than the HE shell and gives a muzzle velocity of 600 m/sec. It penetrates 101 mm of armour at 1,000 m.

The carriage is of box section split trail design with two road wheels. Two castor wheels, which fold over the legs for travelling, are used to help split the trail, and in conjunction with a firing jack beneath the saddle give easy 360° traverse. The jack folds forward to travel clamped beneath the barrel and has an integral lifting device. A spade is permanently fitted to each trail leg, folding up for travelling. In common with other Soviet guns the shield has an irregularly shaped top, is mounted over the wheel axles and overlaps the wheels. A sliding centre section accommodates elevation of the barrel.

A variety of tractors is used including the AT-L, AT-S and URAL 365.

A battalion of three four-gun batteries is organic to Soviet motor rifle divisions.

Employment

Czechoslovakia	India	USSR
Germany (GDR)	Poland	Vietnam
Hungary	Rumania	

152 mm field howitzer M 1955 (D 20) in travelling position

155 mm howitzer

Finland

Calibre	155 mm
Barrel length	33 calibres
Muzzle brake	Single baffle
Trail type	Split
Weight	9,500 kg
Ammunition type	Separate
Charges	—
Ammunition options	HE (43·7 kg)
Rate of fire	—
Muzzle velocity	725 m/sec
Range	20,000 m
Elevation limits	−3° + 52°
Traverse limits	90°
Detachment	10

This weapon shares the same carriage as the Tampella 122 mm M 60. It is made under licence in Israel by SOLTAM and is there known as the M 68. The same gun is fitted to the Israeli L 33 self-propelled equipment,

In action

BAW - D

and it is notable that this towed piece retains the fume extractor used on the SP. The breech mechanism is a horizontal sliding block, and the bore is chrome-plated to resist wear. The split trail carries the wheels, which move as the trail legs are opened.

Employment	Finland
	Israel

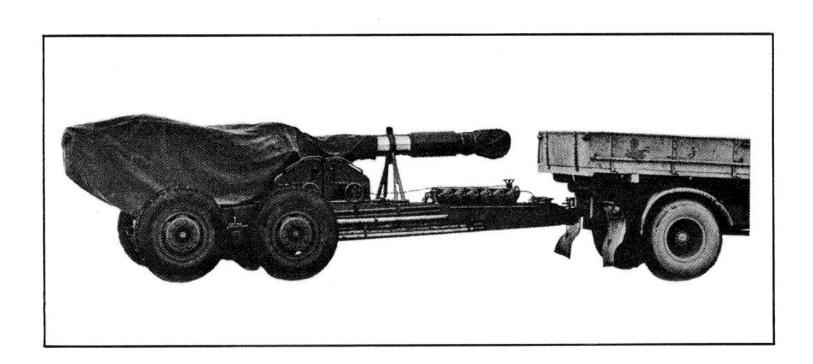

155 mm howitzer in travelling position

155 mm howitzer M 50

France

Calibre	155 mm
Barrel length	23 calibres
Muzzle brake	Multi-baffle
Trail type	Split
Weight	8,200 kg
Ammunition	Separate
Charges	8
Ammunition options	HE (43·75 kg)
	Smoke WP and BE
	Illuminating
Rate of fire	4 rds/min
Muzzle velocity	647 m/sec
Maximum range	17,750 m
Elevation limits	−4° + 69°
Traverse limits	80°
Detachment	11

This equipment is also made under licence in Sweden where it is known as the 'F' field howitzer.

The barrel carries a screw type breech block with self sealing obturation for the bag charges. The multi-baffle muzzle brake is very slim and appears at first to be part of the barrel. The cradle extends rear-wards to form a rest for the loading tray. Nearly horizontal equilibrators run from a massive collar over the trunnions to the forward end of the cradle. The variable length of the recoil system is mounted in the cradle below the barrel.

The gun fires the full range of French 155 mm ammunition (type 56 and TA 68), and other NATO types. The illuminating round is American. A rocket assisted projectile is under development and will increase the range to 35,300 m. Two types of propellant are used in the charge system: charges one to six are fast burning and charges five to eight slow burning.

155 mm M 50 howitzer showing clearly how the wheels swing with the trail legs

Provision is made for split function laying, but normally one layer to the left of the barrel forms all the laying functions.

The split trail carriage is an interesting design. It has spades in the form of plates through which anchor pins are hammered into the ground: this arrangement is doubtless secure but must depend on the terrain for ease of deployment. The plates are removed to the top of the trail for travelling. The tandem road wheels are fixed directly to the trail legs so that they only come parallel for travelling when the trail legs are closed. This layout allows a wide spread of the trail to give a good top traverse—in this case 80°. Elevation is 60° is equally generous, and high angle fire is made possible without preparation by a recoil cut off and a firing jack beneath the cradle to absorb vertical firing stresses.

This equipment is employed in the organic artillery regiments of French mechanised brigades and is to be replaced in the fullness of time by the GCT 155 mm SP.

The weapon has been made under licence in Sweden. Local modifications were made to the trail to speed deployment and increase top traverse to 82°. The only noticeable difference to the appearance of the Swedish gun is the fitting of castor wheels to the trail legs.

Employment France
Israel
Sweden

155 mm howitzer Model 'F'

Sweden

Calibre	155 mm
Barrel length	23 calibres
Muzzle brake	Multi baffle
Trail type	Split
Weight	9,000 kg
Ammunition type	Separate
Charges	8
Ammunition options	HE (43 kg)
Rate of fire	4 rds/min
MV	650 m/sec
Range	17,600 m
Elevation limits	−4° +69°
Traverse limits	80°
Detachment	9

Sweden makes this version of the French 155 mm M 1950 under licence. It is a heavy gun for its calibre and has a complicated system of hydraulics. The heavy spades on the split trails cause it to be slow into and out of action. Israel makes the same gun on a Sherman chassis.

155 mm howitzer Model 'F'

Employment Sweden

155 mm auxiliary-propelled gun 'TR'

France

Calibre	155 mm
Barrel length	40 calibres
Muzzle brake	Multi-baffle
Trail type	Split
Weight	9,700 kg
Ammunition type	Separate loading
Charges	—
Ammunition options	HE (43·25 kg)
Rate of fire	See text
MV	—
Range	24,000 m
Elevation limits	−5° + 66°
Traverse limits	25° left, 38° right
Detachment	8

This weapon has been developed by the Groupement Industriel des Armements Terrestres (GIAT) and is presently in prototype form. It is of very similar type to the Swedish FH 77 and Anglo/German FH 70 in that it adds auxiliary propulsion to a conventional split trail. The APU is in a package ahead of the wheels and drives via an hydrostatic transmission system to give a speed of 4 km/hr. The unit also provides hydraulic power for lifting and lowering the trail dolly wheel, unlimbering, elevating and traversing, and loading. By using the powered loading system a rate of fire of 3 rounds within 15 seconds is claimed, with a sustained rate of 3 rds/minute.

155 mm howitzer FH 70

Tri-national (Federal Republic of Germany, Italy, United Kingdom)

Calibre	155 mm
Barrel length	39 calibres
Muzzle brake	Double baffle
Trail type	Split
Weight	7,800 kg
Ammunition type	Separate
Charges	8
Ammunition options	HE (43·5 kg) Smoke BE Illuminating
Rate of fire	6 rds/min
Muzzle velocity	880 m/sec
Maximum range	24,000 m
Elevation limits	−5° + 70°
Traverse limits	56°
Detachment	10

FH 70 is a joint Italian-German-United Kingdom project. British requirements to replace the 5·5 inch gun, and a German plan to replace its American M 114A1s were reinforced by NATO Basic Military Requirement 39, written in 1963. This stated the need for a close support weapon to be available in towed and SP versions. Germany and Britain carried out joint design studies, and found increasingly similar requirements for a weapon of this kind. By 1970 a towed prototype had started firing trials and Italy joined the project as the third partner. The weapon began to enter service in 1979.

The breakdown of national responsibilities is:

Germany:	The elevating mass Auxiliary Propulsion Unit (APU) Sights Projectiles (less HE)
United Kingdom:	Coordinating authority The carriage HE projectile Propellant system
Italy:	Buffer recuperator and cradle Elevating gear Production assistance for HE shells and some charges

The barrel, of monobloc construction from high yield steel, is autofrettaged at the chamber end. The vertical sliding breech block opens upwards and is closely related to the one fitted to the M 109 G SP howitzer. A ring obturator seals the breech. An automatic ignition tube loader is fitted in the breech block and holds 12 tubes. The muzzle brake is claimed to be 32 per cent efficient.

The recoil system is conventionally under the barrel. Maximum recoil is 1,500 mm: a cut-off gear reduces it at high angles of elevation. The cradle is extended upwards to form the rear mountings for the equilibrators.

Chamber and breech design allow the gun to fire all NATO standard ammunition, but the barrel was designed around its own high performance ammunition.

The thin wall HE shell is claimed to have a better lethality than the American M 107 projectile. Its fuze (P/PD) has been developed to work with a wide range of calibres. A carrier shell provides base ejection smoke and illuminating rounds. Initially a conventional (but new) shell will range to 24,000 m, but rocket assisted (American M 549) projectiles are being developed to increase the range to 30,000 m. The eight charges are divided into three cartridges: one and two, three to seven, and eight.

Three rates of fire are quoted: a burst rate of three rounds in 15 seconds, an initial engagement rate of six rounds/minute, sustained fire at two rounds/minute.

FH 70 ready for firing. Note the forward extension of the carriage to carry the APU and the dolly wheels folded into the crook of the trail legs

Note digital display unit to right of dial sight and loading tray positioned below the breech

Loading is not automatic but 'assisted': during run-out a shell is presented to the breech ready for manual ramming.

The sighting system, which is mounted on the left hand side of the equipment, consists of a dial sight and an anti-tank direct fire telescope. The sight uses the coded disc system of measuring horizontal and vertical angles and this data is transmitted electronically to a data display unit mounted on the equipment. The scales of the sight will be illuminated by nuclear light sources.

The weapon has a split trail, and fires off a turntable sole plate with the wheels raised clear off the ground. A detachable APU provides limited SP mobility. The carriage is extended forward off the wheels to carry a 1,700 cc Volkswagen engine which drives the main wheels. The mechanical couplings are disengaged for conventional towing. Steering is by the trail leg dolly wheels, which are also used to open and close the trail legs on deployment. All other deployment handling is power-assisted. When the APU is detached, power is provided by a hand pump. The main wheels have trailing arm suspension. The suspension's hydro-pneumatic dampers are also used to raise the wheels and thus lower the gun on to its sole plate for firing. The barrel is turned to the rear and clamped over the trail legs for travelling, both when towed and self propelled.

Each country is to produce its own tractor—in all cases a 6 × 6 vehicle.

Employment

User trials for—Italy
Germany (FRG)
UK

FH 70 being moved by APU. The driver controls the gun with conventional foot pedals, and steering is by tiller

155 mm howitzer FH 77A

Calibre	155 mm
Barrel length	38 calibres
Muzzle brake	Pepper pot
Trail type	Split
Weight	11,177 kg
Ammunition type	Separate loading
Charges	6
Ammunition options	HE (42·4 kg) Rocket assisted Smoke Illuminating Practice
Rate of fire	See text
Muzzle velocity	774 m/sec
Maximum range	22,000 m
Elevation limits	−3° + 50°
Traverse limits	60°
Detachment	9

The first ten production guns were introduced into Swedish service in 1978 and supply is currently in progress. The barrel is slightly shorter than that of the FH 70, while the range is somewhat less, due to the weapon being less highly stressed. It is, however, much heavier than the FH 77 due to the addition of mechanical ammunition handing and a more powerful auxiliary propulsion unit and power pack.

Basically the FH 77 is a split-trail weapon of conventional type. The auxiliary power unit, contained in a section ahead of the wheels, is run by a Volvo B20 80 bhp petrol engine; this drives two hydraulic units, one in each gun wheel, allowing the gun to move independently. The APU also supplies pressure to lift or lower the auxiliary support wheels under the trail, allowing power to be used in unhooking the gun from its towing vehicle. The APU can also be controlled from the cab of the towing vehicle, so that it can be brought into play during towing so as to add propulsion to the effort of the tractor. If the travelling speed rises above 8 km/hr the APU is automatically shut off.

The power unit is also used for other operations, in order to reduce crew fatigue and also so that in an emergency the equipment can be operated by one man. An hydraulic crane lifts the ammunition from pallets, three rounds at a time, on to a loading table from which they pass to a power rammer. The cartridge is contained in a steel-based plastic case and shell and cartridge are hydraulically rammed at any elevation. Power is then used to lay the gun, in conjunction with an automated sight developed by Philips Elektronikindustrier AB and Jungner Instrument AB. The sight control unit receives data from the Fire Direction Centre and units within it automatically compute offsets for trunnion dislevel. The actual sight unit is a combined panoramic sight and automatic levelling device; the transmitted firing data automatically sets the sight so that the gunlayer merely has to traverse the gun until his sight is aligned with his aiming mark. He then elevates the gun until two reference marks in the sights coincide. All elevation and traversing is done by hydraulic power, controlled by buttons mounted in hand grips in front of the layer's seat. Once the gun is laid correctly in accordance with the transmitted data, a green light shows on the control unit. The gun is then fired electro-mechanically. In the event of the APU failing, hydraulic power can be maintained by hand pumps.

The HE shell is of a new pattern claimed to give greatly improved lethality; service shells are filled with TNT but it is understood that shells for export will be filled with RDX/TNT mixtures. Base ejection smoke and illuminating shells are also provided, together with a practice shell which has an inert filling plus a small TNT flash charge for indication of impact. A PPD fuze known as 'Zelar' is provided, with seven different options of impact, delay or proximity action

The rate of fire is described as 3 rounds in 8 seconds, followed by 6 rounds every alternate minute for the subsequent 20 minutes. A rocket-assisted projectile is under development which is expected to increase the maximum range to 30,000 m. A super-charge has also been developed, but the Swedish Army have not elected to use this. With the standard HE shell the super-charge would provide 23,500 m and with the RA shell probably 34,000 m.

Employment Sweden

155 mm field howitzer 77 with Scania LBAT 110 cross country truck

155 mm howitzer M 114 A1

United States of America

Calibre	155 mm
Barrel length	20 calibres
Muzzle brake	Nil
Trail type	Split
Weight	5,760 kg
Ammunition type	Separate
Charges	7
Ammunition options	HE (43·2 kg) Nuclear Smoke WP and BE Illuminating Chemical/gas
Rate of fire	1 rd/min
Muzzle velocity	564 m/sec
Maximum range	14,955 m
Elevation limits	−2° + 63°
Traverse limits	49°
Detachment	11

The M 114 entered service as the M 1 in May 1940. It shared the same carriage as the 4·5 inch gun and these two equipments, which complemented each other tactically, were the American medium artillery equivalents of the British 5·5 inch and 4·5 inch guns. Over 6,000 M 114s were produced, and although the range is short for a weapon of this calibre, it is a robust equipment which is still widely used. It has been phased out of service in the US Army.

A variable length hydro-pneumatic recoil system is mounted under and over the barrel. An 'O' ring at the front of the recoil system is also used as the forward mounting for two horizontal spring equilibrators.

The ammunition is widely used in other 155 mm weapons and is a NATO standard type. An initial or burst rate of fire of four rounds/minute is possible: the sustained rate is two rounds/minute.

The split trail legs are girder-type welded box sections. Removable spades are carried on the legs for travelling. Vertical firing stresses are passed on a firing jack which folds forward when not in use. On the M 1 A1 carriage the jack is rack and pinion operated: on the later M 1 A2 it is of the screw type. A small shield is fitted either side of the barrel: its left hand top section folds to enable the dial sight to be used for direct laying.

The weapon is not as heavy as its massive appearance suggests: it can be lifted by a CH 47 class helicopter.

Two variants have been produced:

M 123 A1: This is an SP version using an auxiliary propulsion unit, which is the Continental Motors Corporation four cylinder 20 bhp petrol type motor. It propels the gun at 6–7 km/hr. Drive is to the existing carriage wheels by hydraulic motors mounted outboard. A small dolly wheel assembly supports the trail legs. The modification does not prevent conventional towing, and the firing configuration of the carriage is unaltered. The shield is not fitted. Weight is increased to 6,350 kg. The performance of the gun is unchanged.

Field howitzer FH 155 (L): This variant is being studied by West Germany. The modifications to the weapon by Rheinmetall are still experimental. The carriage is retained, but a new, angled shield changes the weapons appearance. So does the double baffle muzzle brake. It is believed that the barrel and breech ring are new.

Employment

Argentina	Japan	Portugal
Austria	Jordan	Spain
Belgium	South Korea	Taiwan
Brazil	Lebanon	Thailand
Denmark	Laos	Tunisia
Germany (FGR)	Libya	Turkey
Greece	Netherlands	USA
Iran	Pakistan	Vietnam
Israel	Peru	Yugoslavia
Italy	Philippines	

155 mm US M 114 A1 howitzer

155 mm Light howitzer M 198

United States of America

Calibre	155 mm
Barrel length	—
Muzzle brake	Single baffle
Trail type	Split
Weight	6,622 kg
Ammunition type	—
Charges	—
Ammunition options	Nuclear HE
Rate of fire	—
Muzzle velocity	—
Maximum range	30,000 + m
Elevation limits	—
Traverse limits	—
Detachment	11

This weapon is designed to meet the requirement for a highly mobile, air delivered weapon with greater fire power than the 105 mm calibre weapons usually associated with air mobile operations. It has entered service with airborne and air mobile formations this year and is being issued to infantry divisions to replace the M 114 A1.

Although, it weighs 6,622 kg it is considerably lighter than M2 and more robust than the M 114. It can be air lifted by the CH 47C helicopter. Aluminium is used extensively in its construction. The monobloc barrel is autofrettaged to give it strength with lightness. It is mounted at its extreme rear, behind the sliding-wedge breech block. Equilibrators run up from the base of the saddle, but are angled sufficiently to prevent their projection above the recoiling mass. A second pair of equilibrators run foward from a position above the breech. The variable length recoil system is mounted above and either side of the barrel which carried a wedge shaped single baffle muzzle brake. Elevation is by a ball and screw type mechanism.

A new ammunition system is being developed to include nuclear and conventional projectiles. Rocket assisted are to be included, and maximum range is about 30,000 m.

The trail is of deep box section members. A two position suspension setting is employed, the gun being lowered onto a firing platform in action. The platform is carried on the trial legs. Spades are fitted beneath the rear of the trial legs. The weapon can be towed with the barrel deployed, or rotated over the trail legs. The latter position is used for long journeys, and air dropping. No shield is fitted.

Employment USA

Note the extreme rear mounting of the trunions, and balancing springs

180 mm gun S-23

Union of Soviet Socialist Republics

Calibre	180 mm
Barrel length	47 calibres (approx)
Muzzle brake	Pepper pot
Trail type	Split
Weight	21,400 kg
Ammunition type	Separate
Charges	—
Ammunition options	Nuclear HE (88 kg) Rocket-assisted projectile
Rate of fire	1 rd/min
Muzzle velocity	790 m/sec
Maximum range	29,250 m
(with RAP)	43,800 m
Elevation limits	−2° + 50°
Traverse limits	44°
Detachment	16

In the first edition this weapon was described as the **203 mm howitzer M 1955,** but it is now known to be of 180 mm (7 ins), and more appropriately called a gun, with the characteristics of a howitzer (although the distinction between these two types has long since been blurred with the general increase in range and the use of a number of charges for both guns and howitzers) as it has a long gun-like barrel and very long range.

The barrel is supported for over half of its length by a long ring cradle. A square breech ring houses a screw type breech block. The buffer and recuperator cylinders are below the cradle, and the barrel is retracted in the cradle for towing.

A variable bag charge is used, and like the M 115 and M 110 the shell is nuclear or HE. There is also said to be an APHE for use against hard targets, and it would be in conformity with Soviet practice to have a chemical shell as well for such a large capacity projectile.

The box section, split trail carriage, is supported on double road wheels on single axles, and a two wheeled limber supports the trail legs. The AT-T fully tracked tractor is used to tow the gun.

In the Soviet Army, the M 1955 is held at Army or Corps level as a general support weapon.

Employment

Albania	Egypt	Poland
Bulgaria	Germany (GDR)	Rumania
China	Hungary	USSR
Czechoslovakia		

180 mm guns S-32 travelling behind their AT-T tractors

IV. Self-propelled Field Guns

The previous section dealt with towed field guns, and the technical problems of field gunnery and indirect fire. The equipments described in this chapter are employed in the same way as their towed equivalents.

The advantages and disadvantages of SP weapons have been discussed in Section III. In summary, the SP is expensive to develop, manufacture and maintain: it is heavy and therefore difficult to move by air and requires careful route reconnaisance: training is a problem with enclosed fighting compartments: the whole weapon is vulnerable to automotive breakdown. Set against these disadvantages, the SP field gun is more mobile, quicker to deploy and move, often automated, and in some measure independant of an ammunition vehicle. The immunity from direct attack which indirect fire techniques give to field artillery generally is supplemented in the case of most SPs by the protection against all but a direct hit afforded to the detachment and equipment by armoured fighting compartments.

Three types of mounting are found: the open chassis, e.g. M 107, the built-up casemate which offers limited traverse, but is generally more spacious than a turret, e.g. the VK 155 L/50 and the rotating turret, e.g. Abbot.

105 mm SP howitzer AMX 105/50

France

Calibre	105 mm
Barrel length	23 calibres
Muzzle brake	Double baffle
Ammunition type	Separate
Charges	8
Ammunition options	HE (Mk 63) (16·15 kg) HE (M 1) (14·9 kg) HESH Smoke Chemical Illuminating Carrier Blank
Rate of fire	—
Muzzle velocity	568 m/sec (Mk 63 ammunition)
Maximum range	14,000 m (,, ,,)
Elevation limits	−4°30′ + 70°
Traverse on carriage	20°
Detachment	4
Chassis—Type	AMX 13
Engine	SOLAM, flat 8, petrol
Power	270 bhp
Speed	60 km/hour
Endurance	—
Ammunition carried	52
Height × length × width	2·7 × 5·2 × 2·65 m
Weight	16,500 kg

This weapon is also referred to as the AMX 105/A. It entered service in 1952 and is still used in the artillery regiments of non-mechanised brigades and in the divisional artillery reserve units.

The gun was designed to use the standard American M 1 ammunition which gives a maximum range of 11,500 m. This performance has been improved by the introduction of the French Mk 63 ammunition: the HE shell is of a hollow base design and the new HEAT round penetrates 360 mm of armour. A new double charge system comes in two cartridge cases—one to five and six to eight. The claimed burst fire rate is six rounds in 20 seconds.

The weapon's chief limitation is the poor top traverse due to the fixed casemate mounting. A distinctive recognition feature is the square shield, drilled in the top left corner for the direct fire telescope's line of sight, which elevates with the recoiling mass.

The chassis is a derivative of the AMX 13 light tank. It provides good cross country mobility and a stable gun platform for this size of gun so that a spade is not needed. This contrasts with the extensive changes necessary to mount a 155 mm on this chassis (see entry relating to 155 mm SP Mk III).

The Dutch AMX 105/50 is fitted with a longer 30 calibre barrel. This barrel, which increases range to 15,000 m is also fitted in the newer 105 B French gun.

Employment France
Morocco
Netherlands

The weapon is mounted in a fixed casemate. Note the shaped armour covering for the buffer/recuperator beneath the barrel, and the unusual shield with its aperture for direct laying of line of sight

105 mm SP howitzer AMX 105B

France

Calibre	105 mm
Barrel length	30 calibres
Muzzle brake	Double baffle
Ammunition type	Separate
Charges	8
Ammunition options	HE (16·15 kg) Smoke Illuminating
Rate of fire	—
Muzzle velocity	675 m/sec
Maximum range	15,000 m
Elevation limits	—
Traverse on carriage	360°
Detachment	4
Chassis—Type	AMX 13
Engine	SOLAM, flat 8, petrol
Power	270 bhp
Speed	60 km/hour
Endurance	—
Ammunition carried	80
Height × length × width	2·7 × 6·4 × 2·5 m
Weight	17,000 kg

The AMX 105B is an improved version of the AMX 105/50. The main difference lies in the provision of a turret giving 360° traverse: because the 105/50 already has a fully enclosed casemate this new equipment, with turret, is only 500 kg heavier. A commander's cupola is fitted, and mounts a machine gun.

The longer barrel, as fitted to the Dutch 105/50 is used with the original muzzle brake. The recoil and equilibrator systems are housed inside the turret.

The ammunition system is the eight charge Mk 63 type used in the 105/50, but performance is improved to 15,000 m by the longer barrel. The vehicle carries 80 complete rounds of ammunition which is an important improvement over the 105/50. It is not known if the new barrel accepts American M 1 ammunition.

Chassis details are as for the 105/50.

Employment Switzerland

105 mm SP gun 'Abbot'—FV433

United Kingdom

Calibre	105 mm
Barrel length	325 cm
Muzzle brake	Double baffle
Ammunition type	Semi-fixed
Charges	8
Ammunition options	HE HESH Smoke Illuminating Practice
Rate of fire	12 rds/min
Muzzle velocity	705 m/sec
Maximum range	17,300 m
Elevation limits	−5° + 70°
Traverse on carriage	360°
Detachment	4 (including driver)
Chassis—Type	FV 430
Engine	Rolls Royce K60
Power	240 bhp
Speed	48 km/hour
Endurance	480 km
Ammunition carried	40
Height × length × width	2·5 × 5·8 × 2·6 m
Weight	17,463 kg

Abbot is the first British gun to be designed as an SP from its inception. The barrel is new and the chassis is a derivative of the APC FV 430 series. It came into service in 1965 as the standard close support weapon of the Royal Artillery, except in specialist and airportable functions, replacing the towed 25 pounder.

The gun's performance is exceptional. With the latest Mk II ammunition, maximum range is 17,300 m. Accuracy at that range is twice that of previous designs and barrel life is at least 10,000 rounds.

A semi-automatic loading system with powered ramming produces a sustained rate of fire of 12 rounds/minute. The spent cartridge is ejected automatically on run-out and the vertical sliding breech closes when a fresh cartridge is loaded. The cartridge is brass, and is fired electrically. Eight charges are available. Cartridges are supplied with five charge bags. These are removed and the cartridge re-made for sub charges A and B. Super charge is delivered in a special cartridge case.

The gun accepts American M 1 ammunition using Abbot Mk I charges (range limited to 15,000 m), and its own range of HE, HESH, SH practice, illuminating and smoke shells. A nylon driving band is under development to provide exceptional consistency and increased range while reducing mechanical wear to a minimum.

The turret rotates through 360° and provides protection against small arms fire and shell splinters. The crew is protected against chemical attack and nuclear fall out. The pressurized crew compartment is air-conditioned, and a fume extractor is fitted to the barrel.

105 mm SP gun—'Abbot'

The Abbot prepared for swimming, with the flotation screen erected and the exhaust pipe extension fitted

Sighting is conventional and laying is electrically power assisted. Fire control is effected by a sophisticated communications kit. Communication within the turret is by an induction loop feeding special head sets. A subsidiary loop communicates outside the vehicle. Radio and sighting systems are compatible with developments in automatic data transmission.

The Rolls Royce K60 engine is a turbo-charged vertically opposed six cylinder multi-fuel type. Its 240 bhp gives a power to weight ratio of 15 bhp/ton. The Allison Tx 200 automatic transmission is fitted. Electrical power is provided by two alternators running off the main engine to 24 volt batteries of 200 amp/hour capacity.

The road wheels have torsion bar suspension. No spade is fitted. Ground pressure is 0·81 kg/cm². The vehicle wades to four feet without preparation, and swims propelled by its tracks when a permanently fitted flotation screen is raised.

A less sophisticated 'value engineered' version is made by the manufacturer (Vickers Ltd) called the Vickers Abbot. It is recognisable by the absence of external stowage bins. Models now in production are fitted with the General Motors V6 petrol engine which powers the American M 113 APC.

The Vickers private venture Falcon twin 30 mm air defence gun is mounted on the Abbot chassis.

Employment India
UK

105 mm SP howitzer M 52

United States of America

Calibre	105 mm
Barrel length	—
Muzzle brake	Nil
Ammunition type	Semi-fixed
Charges	7
Ammunition options	HE (14·9 kg) HEAT Smoke WP Chemical Illuminating Carrier Blank
Rate of fire	3 rds/min
Muzzle velocity	472 m/sec
Maximum range	11,200 m
Elevation limits	−10° + 65°
Traverse on carriage	120°
Detachment	5
Chassis—Type	M 41 light tank
Engine	OAS 895
Power	500 bhp
Speed	56 km/hour
Endurance	150 km
Ammunition carried	105
Height × length × width	3·1 × 5·8 × 3·3 m
Weight	24,040 kg

The M 52 was introduced into service in 1954, part of a new family of SP equipments based on the chassis of the M 41 tank. Previous American SPs had been open mountings, but in view of the nuclear fall-out danger, the M 52 had a partially-rotating turret. A notable feature is the location of the driver in the turret.

The gun was designed to use the M 1 ammunition which was originally intended for the M 101 towed howitzer. The breech is a vertical sliding type. No muzzle brake is fitted, but a long slim fume extractor is fitted.

The chassis used for the M 52 was first developed for the 155 mm M 44. The use of a smaller gun greatly increases the stowage available for ammunition, and does away with the need for a spade. The chassis is fully described under the M 44.

Employment Japan, Jordan, Spain, Tunisia, USA (Reserve)

105 mm SP howitzer M 52

105 mm SP howitzer M 108

United States of America

Calibre	105 mm
Barrel length	—
Nuzzle brake	Nil
Ammunition type	Semi-fixed
Charges	7
Ammunition options	HE Illuminating Smoke Chemical
Rate of fire	3 rds/min
Muzzle velocity	473 m/sec
Maximum range	12,000 m
Elevation limits	−4° + 74°
Traverse on carriage	360°
Detachment	7
Chassis—Type	Individual
Engine	General Motors 8V 71 T
Power	420 bhp
Speed	55 km/hour
Endurance	350 km
Ammunition carried	87
Height × length × width	3·15 × 6·09 × 3·29 m
Weight	22, 452 kg

The 105 mm M 108 was introduced in 1964. It consists of the M 52 barrel mounted on a special chassis which it shares with the 155 mm M 109.

The barrel is not fitted with a muzzle brake, but has a small fume extractor near the muzzle. The ammunition is the M 1 type developed for the M 101A1, but unlike that weapon, the M 108 has a vertical sliding breech inherited from the M 52. The power assisted loading system is shared with the larger M 109, and results in a low rate of fire.

Sighting is by conventional dial sight and elevation quadrant. There is a direct fire elbow periscope. Laying in elevation and azimuth is hydraulically powered.

Extensive use of aluminium keeps the weight down. The turret gives protection to the detachment against small arms fire, and splinters. The turret is very spacious, with good access from side and rear hatches.

The chassis runs on 14 road wheels, using torsion bar suspension. No return rollers are used for the 15 inch wide track. The engine is the General Motors 8V 71 T turbo-charged diesel delivering 420 bhp through an Allison XTG 411-2A automatic transmission. There are four gears: steering is by clutch brake on the lower two, and by geared steering on the higher ratios. Power for the hydraulic firing functions is provided by a pump driven off the main engine. As a back-up service there is a manual pump for elevation and traversing hand wheels.

The vehicle can swim when fitted with flotation bags in canvas retainers. Four are fitted each side, and one at either end. Wave barriers at side and front are also fitted. A blower inflates the bags in 75 seconds to two pounds/square inch. This pressure can be maintained in the event of minor damage. Propulsion in the water is by track motion at a speed of 6 km/hr.

Employment

Belgium	Sweden	Turkey
Brazil	Switzerland	USA
Israel	Taiwan	Yugoslavia
Spain	Tunisia	

105 mm SP howitzer M 108 in action in Vietnam

122 mm SP howitzer

Union of Soviet Socialist Republics

Calibre	122 mm
Barrel length	47 calibres
Muzzle brake	Double baffle
Ammunition type	Separate
Charges	—
Ammunition options	HE (25·5 kg) APHE Smoke Illuminating
Rate of fire	—
Muzzle velocity	900 m/sec
Maximum range	21,900 m
Elevation limits	—
Traverse on carriage	360°
Detachment	—
Chassis—Type	PT 76
Engine	2 × 6 cylinder Otto
Power	480 bhp
Speed	44 km/hour
Endurance	—
Ammunition carried	—
Height × length × width	Width 3·14 m
Weight	—

This is a new 122 mm SP gun which was seen for the first time at the 30th Anniversary of the Polish Peoples Republic celebrations in 1974. It is the latest Soviet SP howitzer. As yet little is known, and much of the detail quoted is speculative. It is the D 30 gun in a 360° turret, mounted on a new chassis, the front of which resembles the BMP. The driver sits to the left of the engine compartment.

The general layout of the running gear is that of the PT 76 tank with an extra wheel, making a total of 7 per side.

The D 30 gun carries the same double-baffle muzzle brake (and is assumed to give the same performance) as the towed version. Range would be 15,200 m with a 22 kg shell and an anti-armour capability to defeat 230 mm of plate at 1000 m. A rocket assisted round has been rumoured.

With the PT 76 chassis this weapon shares that tank's inherent buoyancy and in-water speed of 10 kilometres propelled by two hydro-jets. The two six-cylinder 240 bhp engines drive through a five-speed gearbox and give a speed of 44 km/hr. Suspension is by torsion bars with two hydraulic dampers per side.

Employment
Germany (GDR) Poland USSR

This picture was taken at the gun's first public showing in July 1974, at the parade to celebrate the 30th anniversary of Poland's liberation

122 mm SP gun M 1974: the similarity of the hull to that of the PT 76 tank is very clear

152 mm SP howitzer

Union of Soviet Socialist Republics

This equipment represents the next logical step after the 122 mm SP howitzer in the Soviet policy of converting its divisional artillery into SPs, with all the advantages in both conventional and nuclear warfare conferred by tracks and an armoured, fully enclosed fighting compartment for the crew. Following the usual economical Soviet practice, a well proved weapon has been married to a well proved chassis. No details are known, but it can be safely assumed that the performance of the gun is the same as the M 1955 (D20)—*see* page 49. The chassis is believed to be the adapted SA-4 (page 217).

152 mm SP howitzer

155 mm SP howitzer Mk III

France

Calibre	155 mm
Barrel length	33 calibres
Muzzle brake	Double baffle
Ammunition type	Separate
Charges	10
Ammunition options	HE (43·25 kg) Smoke BE and WP Illluminating
Rate of fire	4 rds/min
Muzzle velocity	765 m/sec
Maximum range	21,600 m
Elevation limits	0 + 67°
Traverse on carriage	50°
Detachment	11
Chassis—Type	AMX 13
Engine	SOLAM, flat 8, petrol
Power	270 bhp
Speed	60 km/hour
Endurance	—
Ammunition carried	Nil
Height × length × width	2·12 × 6·22 × 2·7 m
Weight	16,700 kg

This equipment consists of the AMX 13 chassis on which is mounted a modified 155 M 50 howitzer. Barrel length has been increased by 10 calibres to 33, and a newer double baffle muzzle brake fitted. Some early models were fitted with the un-modified gun.

The breech is a screw type, with integral obturation for the bag type charge system. Conventional laying arrangements are on the left of the barrel, and the buffer/recuperator beneath it.

The effect of the lengthened barrel is to increase muzzle velocity to 765 m/sec and range to 21,600 m using the TA 68 hollow-base shell. This equipment accepts two charges (9 and 9A) more than the M 50, so that its performance with older types of ammunition is better.

The gun fires French 155 mm ammunition (Type 56 and TA 68) as well as NATO types—including the American M 107 which gives a limited range of 18,000 m. The illuminating round is American. Two types of propellant are used in the charge system: charges one to six are fast burning, and charges five to 9A slow burning.

The gun is large for the size and weight of its chassis, and is therefore mounted at its extreme rear, so that the firing stresses can be passed directly to the ground via two spades which form a miniature split trail. This arrangement allows 50° top traverse, (30° to right and 20° to left): laying in azimuth is power assisted. The positioning of the breech provides excellent access for loading—a task which is possible at the highest elevation (67°). However, the physical effort of loading the separate ammunition keeps the rate of fire to four rounds/minute—as for the towed version.

The equipment does not swim but wades to a depth of 0·7 m. To accommodate the gun and spade mounting the rear sprockets of the tank running gear have been removed, leaving five bogie wheels and the front driving sprocket. Otherwise the layout and performance of the chassis is unchanged.

Normally two crew members (driver and commander) travel with the gun in protected compartments. The commander has a small cupola. Three other members of the detachment can be carried for short distances, but they will normally travel with the remainder in the support vehicle with the first-line ammunition (this second vehicle is the AMX 13 VCA).

A radio is fitted and can be used for fire orders.

Employment Argentina
France
Venezuela

The 155 mm M 50 howitzer mounted on the AMX 13 tank chassis. The trunnions are mounted at the extreme rear over the spade system. Note absence of crew space

155 mm GCT SP gun

France

Calibre	155 mm
Barrel length	40 calibres
Muzzle brake	Multi-baffle
Ammunition type	Separate (combustible cartridge)
Charges	7
Ammunition options	HE (43·75) Smoke BE and WP Illuminating
Rate of fire	9 rds/min
Muzzle velocity	810 m/sec
Maximum range	23,500 m (with TA 68)
Elevation limits	−5° + 66°
Traverse on carriages	360°
Detachment	4
Chassis—Type	AMX 30C
Engine	Hispano-HS 110
Power	720 bhp
Speed	60 km/hour
Endurance	450 km
Ammunition carried	42
Height × length × width	3·0 × 10·4 × 3·1 m
Weight	43,000 kg

The GCT gun (Grand Cadence de Tir) is a new weapon mounted on the well tried AMX 30C tank chassis. It was first seen in 1973.

The long, 40 calibre, barrel improves the range of the Mk III SP howitzer. A new, vertical sliding breech block is fitted. Recoil is constant at 0·95 m despite elevation to 66°. At some traverse angles, elevation is limited to 12°, but not in the main forward arc of fire.

The barrel is chambered to accept existing ammunition such as the American M 107 and French Type 56. However, it is particularly designed to fire the new hollow base TA 68 shell. The charge is contained in a combustible cartridge case. The virtue of this system is that the facility for automated handling and loading, which is not possible with a bag charge, does not lead to the problem of disposing of a spent cartridge case. Older bag charges can be used, but not automatically loaded.

A rocket assisted projectile will increase range to 30,000 m.

42 complete rounds are carried in the vehicle. They are loaded through wide-opening doors in the rear of the turret into a magazine. From the magazine, loading by a hydraulic mechanism is fully automatic. The cyclic rate of fire is nine rounds/minute, which gives a burst fire rate of six rounds in 40 seconds. Manual loading limits the rate to 2–3 rounds/minute.

The AMX 30 chassis is powered by the Hispano-Suiza HS 110 engine. It is a flat 12, multi-fuel supercharged diesel developing 720 bhp, or 18 bhp/ton. The five speed gearbox has a manual shift. The centrifugal clutch is electrically operated by the movement of the gear lever. Five double-wheels per track have torsion bar with resilient bump stops. Hydraulic dampers are fitted to the suspension of the front and rear wheels.

The vehicle wades to 2·2 m, but does not swim.

The equipment is scheduled for service in 1978.

The GCT gun and turret has been fitted to the German Leopard tank chassis as a private cooperative venture between the manufacturers of these two weapons.

Employment France

The great size of the turret of this equipment, mounted on a main battle tank chassis (AMX 30), is apparent from this picture. Although the gun is taken from the earlier Mk III SP, the only noticeable similarity is the muzzle brake

155 mm SP howitzer L 33

Israel

Calibre	155 mm
Barrel length	33 calibres
Muzzle brake	Single baffle
Ammunition type	Separate
Charges	8
Ammunition options	HE (43·7 kg) Smoke Illuminating
Rate of fire	—
Muzzle velocity	725 m/sec
Maximum range	21,525 m
Elevation limits	−3° + 52°
Traverse on carriage	60°
Detachment	8
Chassis—Type	M 4 Super Sherman
Engine	Cummins diesel
Power	460 bhp
Speed	36 km/hour
Endurance	260 km
Ammunition carried	54
Height × length × width	3·46 × 8·55 × 3·33 m
Weight	42,250 kg

The L 33 is the first heavy artillery weapon to be designed and built in Israel by SOLTAM. It consists of the Tampella designed M 68 howitzer mounted on a modified M 4 Super-Sherman tank chassis.

The barrel is fitted with a semi-automatic breech, and this in conjunction with a pneumatic handling and loading system allows a 'very high' rate of fire; no figures have been released. The barrel is chromium plated to prolong its life: replating is possible and restores a worn barrel to a new condition. The range of the weapon is believed to be greater than the published figure.

The gun is mounted in a fixed casemate which forms a high superstructure on the chassis. The recoil system, mounted below the barrel, and the equilibrators, extend beyond the casemate. A fume extractor is fitted.

The aiming system includes direct fire sights.

The Super Sherman tank has been in Israel's service since 1950. Local modifications included the fitting of a Cummins diesel engine. The original horizontal volute suspension is retained. There is little doubt that as a tank this is no longer a viable weapon, and that the serviceable chassis are being put to good use as SP artillery carriages. The crew of eight are served with an induction-loop internal communications system, and are all provided with seat belts.

It is claimed that fuel and ammunition for 20 hours is carried in the vehicle. The weapon is used extensively in the counter-battery role, and considerable thought has been given to its tactical handling. The sighting and fire control techniques have been developed to take advantage of the gun's mobility to change gun positions rapidly, and so avoid enemy counter-battery fire.

Employment Israel

The hull of the original Sherman tank can be seen beneath the new superstructure

155 mm gun SP 70

Tri-national: Federal Republic of Germany, Italy, United Kingdom

Calibre	155 mm
Barrel length	39 calibres
Muzzle brake	Double baffle
Ammunition type	Separate
Charges	8
Ammunition options	HE (43·5 kg) Smoke BE Illuminating
Rate of fire	—
Muzzle velocity	827 m/sec
Maximum range	24,000 m
Elevation limits	—
Traverse on carriage	360°
Detachment	6
Chassis—Type	individual
Engine	—
Power	—
Speed	—
Endurance	—
Ammunition carried	—
Height × length × width	—
Weight	—

SP 70 is a project being undertaken by Germany, Italy and the United Kingdom in parallel with the FH 70 towed gun project. SP 70 will be the FH 70 ordnance mounted on a chassis developed from the successful German 'Leopard' Main Battle Tank. The breakdown of development work is:

Italy	Recoil system Loading system Some chassis elements
Germany	Chassis and running gear Ordnance
United Kingdom	Turret Sights

Performance will be directly comparable with FH 70. Existing 155 mm ammunition and a purpose-built type (HE, smoke and illuminating), will be used. The rocket assisted projectiles which are under development will increase range to 30,000 m. Further details will be found under FH 70.

Employment Under development:
Germany (FRG)
Italy
UK

155 mm gun SP 70 traversed full left

155 mm automatic gun VK 155 L/50

Sweden

Calibre	155 mm	
Barrel length	—	
Muzzle brake	Pepper pot	
Ammunition type	Semi-fixed	
Charges	—	
Ammunition options	HE (48 kg)	
Rate of fire	15 rds/min	
Muzzle velocity	865 m/sec	
Maximum range	25,000 m	
Elevation limits	−3° + 40°	
Traverse on carriage	30°	
Detachment	4	
Chassis—Type	Individual	
Engine	Rolls Royce K60	Boeing 502–10MA
Power	240 bhp	300 bhp
Speed	35 km/hour	
Endurance	—	
Ammunition carried	14	
Height × length × width	3·25 × — × 3·3 m	
Weight	51,000 kg	

The requirement for a gun of long range was stated in 1950, the prototype appeared in 1960 and production was between 1966 and 1968. This highly advanced Swedish SP from AB Bofors features fully automatic firing at a rate of 15 rounds/minute. To achieve this remarkable performance the all-round traverse featured in most modern SP designs has been sacrificed: only 30° of traverse is possible without moving the vehicle itself: there is no spade to delay this action.

The magazine holds 14 rounds. The ammunition is delivered to the gun in clips of five rounds, which are loaded into the magazine mechanically. Fuze and charge options are selected before the magazine is loaded.

Recoil forces are used to cock the spring action of the loading system which consists of two movements: offering the round to the breech, and ramming. From the magazine, rounds are fed by gravity into the two feed trays. The feed trays serve a loading tray behind the breech.

Range is very good—25,000 m, and it is believed that a rocket assisted projectile is being developed which should produce at least 30,000 m.

Laying is manual except that electrical power is used for coarse elevation setting, and for raising and lowering the barrel to and from the loading angle.

All automatic and powered functions have manual back-up systems, for which the detachment in the two cabins is increased to seven. The vehicle is enormously heavy. Two engines are fitted: Rolls Royce K60 (as fitted in Abbot) is a turbo-charged six cylinder two stroke diesel producing 240 bhp: a Boeing 502-10MA gas turbine plant produces 300 bhp. The vehicle runs on 12 road wheels and the track returns along the top of these without rollers. The hydro-pneumatic suspension is locked for firing.

22 mm of armour plate is fitted to the magazine and cabin which is fully proofed to protect the crew from chemical attack and nuclear radiation.

Employment Sweden

155 mm SP gun VK 155 L/50: note the massive magazine mounted over the turret

155 mm SP howitzer M 44

United States of America

Calibre	155 mm
Barrel length	20 calibres
Muzzle brake	Nil
Ammunition type	Separate
Charges	7
Ammunition options	HE (43·2 kg) Nuclear Smoke WP and BE Illuminating Chemical
Rate of fire	1 rd/min
Muzzle velocity	569 m/sec
Maximum range	14,600 m
Elevation limits	−5° + 65°
Traverse on carriage	60°
Detachment	5
Chassis—Type	M 41 light tank
Engine	OAS-895
Power	500 bhp
Speed	56 km/hour
Endurance	120 km
Ammunition carried	24
Height × length × width	3·1 × 6·09 × 3·3 m
Weight	28,350 kg

The M 44 is now an outdated equipment, relegated to the reserve by the United States' Army, and largely phased out by previous users. It was the last of the American SP guns to have an open topped casemate.

The howitzer is the M 114, but with a modified recoil system which does not project beyond the casemate. The interrupted screw type breech block and percussion firing system are retained.

The ammunition used is the M 107 family which was developed for the M 114. A spring ram is fitted.

The gun is mounted in an open platform with high sides but no top. Traverse is good for the style of mounting. Stability in action comes from a large, hydraulically operated spade, over which the rear of the casemate opens to extend the platform floor, and ease ammunition supply.

The chassis is based on the M 41 light tank, and is also used for the M 52 SP. The running gear has six road wheels, four return rollers, and a single (driving) sprocket. The engine is a supercharged horizontally opposed six cylinder, and is air cooled. An unusual feature is the driving position which is to the left of the breech in the casemate. The engine and transmission fill the whole of the front of the hull.

The main weaknesses in the design are the lack of overhead protection for the detachment, the small on-board ammunition store and the limited radius of action.

Employment		
	Belgium	Japan
	Ecuador	Jordan
	Greece	Spain
	Israel	Turkey
	Italy	USA (Reserve Army)

155 mm SP howitzer M 44: note rear doors open, extended platform and hydraulic spade at rear

155 mm SP howitzer M 109

United States of America

Calibre	155 mm
Barrel length	20 calibres
Muzzle brake	Double baffle
Ammunition type	Separate
Charges	7
Ammunition options	Nuclear HE (43·2 kg) Smoke BE and WP Illuminating Chemical Cannister
Rate of fire	45 rds/hr
Muzzle velocity	561 m/sec
Maximum range	14,600 m
Elevation limits	$-3° + 75°$
Traverse on carriage	360°
Detachment	6
Chassis—Type	Individual
Engine	General Motors 8V 71 T
Power	420 bhp
Speed	55 km/hour
Endurance	355 km
Ammunition carried	28
Height × length × width	3·05 × 6·62 × 3·15 m
Weight	23,769 kg

The M 109 has become a standard workhorse for general support in the western world. Its size, spacious turret and reliable chassis make it an ideal vehicle for development of other projects. The M 109G and M 109U are national variants developed by Germany and Switzerland respectively. The M 108 is a 105 mm version, and the M 109 A1 a long barrel development now entering service as a retrofit conversion. The Italians have modified one as a development vehicle for SP 70.

In its original form the M 109 range is only 14,600 m. However, within that limitation its charge system gives excellent overlap and high angle fire is instantly available. The breech is interrupted screw type with integral obturation for the bagged charge. The firing mechanism is percussion using a lanyard. A hydraulic, semi-automatic power ram operates at a set loading angle. A loading tray and the rammer are deployed from the rear of the turret for each operation. Although elevation is fully powered the rate of fire is 45 rounds/hour.

The M 114 ammunition is used, but with the addition of a cannister round. The Americans are known to have developed a 155 mm nuclear shell which may be fired from this weapon.

Sighting is by conventional dial sight and elevation quadrant. There is a direct fire elbow periscope. Hydraulic power rotates the turret and is used for elevation. Power for the hydraulic firing functions is provided by a pump driven off the main engine. As a back-up service there is a manual pump for elevation and traversing.

The M 109 has the same chassis and turret as the 105 mm M 108. It can be distinguished by the heavier barrel and bulky six valve fume extractor mounted close to the muzzle brake. The M 109 also has two small spades behind the tracks for additional stability on reverse slopes and when firing high charges at low angles of elevation.

The chassis runs on 14 road wheels, using torsion bar suspension. Non-return rollers are used for the 15 inch

155 mm SP howitzer M 109

This picture shows the small spades deployed and gives an impression of the spaciousness of the turret. Note the side access panel to the fighting compartment

The interior of the German M 109G showing the horizontal sliding breech and, to its left, the Zeiss periscopic sight

wide track. The engine is turbo-charged diesel delivering 420 bhp through an Allison XTG 411-2A Automatic transmission. There are four gears: steering is by clutch brake on the lower two, and by geared steering on the higher ratios. The power to weight ratio of 18 bhp/ton gives the vehicle a good performance.

The vehicle can swim when fitted with flotation bags in canvas retainers. Four are fitted each side, and one at either end. Wave barriers at side and front are also fitted. A blower inflates the bags in 75 seconds to two pounds/square inch. This pressure can be maintained in the event of minor damage. Propulsion in the water is by track motion at a speed of 6 km/hr. The three variants of the basic design are:

M 109 A1: This equipment is the standard M 109, rebarrelled, and using a new charge system. The re-barrelling programme has started in America and the conversion is available to other users. The retention of the fume extractor keeps the original look of the equipment. The original shell is used. The new barrel is 33 calibres in length and raises the muzzle velocity to 635 m/sec. Range is now 20,575 m, while the weight of the equipment has risen to 24,040 kg.

M 109G: This is a German modification which gives a range of 18,000 m—a performance which predates and pre-empts the fitting of the M 109 A1 barrel. The modifications lead to an increase in the rate of fire to six rounds/minute, but this increase is largely due to the fact that the German gunners dispense with the assisted loading system.
Modifications are:

(a) New muzzle brake.
(b) New breech ring and block. The screw type is replaced by a horizontal sliding semi-automatic type. The chamber accepts existing ammunition.
(c) The addition of an eighth charge giving a range of 18,000 m with a muzzle velocity of 686 m/sec.
(d) Locally manufactured Zeiss sights.
(e) Secondary armament replaced by the German MG 1 (formerly known as MG 42).
(f) New track with removable rubber pads as used by the Leopard tanks.

M 109U: This Swiss version is known locally as the Panzerhaubitz 66. It replaces a towed 105 mm howitzer as the artillery weapon organic to the mechanized division. Modifications are:

(a) A new electrical system
(b) A modified loading system which operates at all angles of elevation
(c) The 0·50 machine gun is replaced by a 12·7 mm anti-aircraft gun.

The new long-barrelled M 109 A1

Employment

Argentine	Greece	Norway
Australia	Iran	Pakistan
Austria	Israel	Peru
Belgium	Italy	South Africa
Canada	Jordan	Spain
Denmark	Kampuchea	Switzerland (M 109U)
Ecuador	Libya	UK
Ethiopia	Morocco	USA (M 109A1)
Germany (FRG) (M 109G)	Netherlands	

175 mm SP gun M 107

United States of America

Calibre	175 mm
Barrel length	60 calibres
Muzzle brake	Nil
Ammunition type	Separate
Charges	3
Ammunition options	HE (66·6 kg) Chemical
Rate of fire	30 rds/hr
Muzzle velocity	923 m/sec
Maximum range	32,700 m
Elevation limits	+2° + 65°
Traverse on carriage	60°
Detachment	13
Chassis—Type	Individual
Engine	General Motors 8V 71 T
Power	420 bhp
Speed	54 km/hour
Endurance	725 km
Ammunition carried	1
Height × length × width	3·47 × 11·3 × 3·14 m
Weight	28,165 kg

This is a high velocity gun of considerable range, which entered service in 1962. It has served as an important counter-battery weapon in NATO and is fully air-portable. At the time of writing it is being phased out of British and US service, to be replaced by the 8 inch M 110 A1 howitzer, since this new barrel will interchange with the 175 mm barrel. It is expected that all 175 mm guns in US and British service will have been converted by mid-1980.

The performance stems from the great length of barrel, but at the cost of a barrel life of only 1,200 EFC's. The interrupted screw type breech block provides obturation for the bag charge system. A variable length hydro-pneumatic recoil system is mounted below the barrel.

Split function laying is used. The azimuth layer sits to the left of the barrel, the elevation layer to the right.

Loading is fully mechanized: the hydraulic systems are controlled from the platform behind the breech. The shell is delivered to the gun in a carrying tray. A hoist lifts the carrier and round to a loading tray which is offered to the breech for power ramming. One round is carried on the equipment.

The vehicle layout is non-tactical in that the chassis is open and the gun mounted without casement or turret: thus the overall weight of the weapon is kept within reasonable limits. Of the detachment of 13, five travel on the equipment. The driver sits down in the chassis and is offered some protection by a hatch. The massive spade is raised and lowered hydraulically. Stability in action is assisted by locking the road wheel suspension. The barrel is retracted for travelling.

The running gear consists of five large diameter road wheels per side: the track returns along their upper rims. Ground pressure is 0·95 kg/cm². Hydraulic power is produced by a pump run off the main engine—a super-charged diesel which is common to the M 108 and M 109.

Employment

Germany (FRG)	Italy	UK
Greece	Netherlands	USA
Iran	Spain	Vietnam
Israel	Turkey	

175 mm SP gun M 107 showing chassis details

The same gun in action: notice spades embedded

8 inch SP howitzer M 110

United States of America

Calibre	203 mm
Barrel length	25 calibres
Muzzle brake	Nil
Ammunition type	Separate
Charges	7
Ammunition type	Nuclear HE (90·7 kg) HE spotting Chemical
Rate of fire	½ rd/min
Muzzle velocity	594 m/sec
Maximum range	16,800 m
Elevation limits	$-2° + 65°$
Traverse on carriage	60°
Detachment	13
Chassis—Type	Individual
Engine	General Motors 8V 71 T
Power	420 bhp
Speed	54 km/hour
Endurance	725 km
Ammunition carried	1
Height × length × width	2·6 × 7·48 × 3·14 m
Weight	26,355 kg

The M 110 consists of the M 115 towed 8 inch ordnance on the 175 mm SP mounting of the M 107. A full description of the two parts will be found under those equipments. The complete interchangeability of the 175 mm and 203 mm barrels is demonstrated by the Israeli M 107 batteries which carry 8 inch barrels, to be quickly interchanged for precision shoots, or long range engagements.

The barrel has an interrupted screw type breech with a variable length recoil system mounted below it. Details of the nuclear round are classified. Two charge systems are used—charges one to five green bag and five to seven white bag. The weapon is noted for its extreme accuracy.

Firing functions are fully power assisted, including the loading sequence. Split function laying is used.

The chassis is powered by a General Motors supercharged diesel engine. Hydraulic power is supplied by a pump run off the main engine. The suspension of the five road wheels per side is locked for firing, but the weapon's stability stems mainly from the huge hydraulically powered spade.

A new long-barrelled version, the M 110 A1, is currently replacing the M 110 and M 107 175 mm gun in British and US service. The new barrel is 40 calibres long and chrome-lined. A new cartridge system, with eight charges, increases the range to 20,600 m, the maximum muzzle velocity now being 710 m/sec.

The M 110 A1 is considered to be an interim conversion measure; it will, in due course, be succeeded by a completely new weapon the M 110 A2. This is, in effect, the M 110 A1 with the addition of a double-baffle muzzle brake. This will permit firing a completely new cartridge system of nine charges plus a completely new family of projectiles, giving a maximum muzzle velocity of approximately 750 m/sec and a maximum range the order of 22 km with a conventional shell. To extend the range the XM 650 rocket-assisted projectile is being developed, which is expected to range to 29,100 m. In addition, the XM 753 rocket - assisted nuclear

The long-barrelled 8 inch SP M 110 (A1)

projectile will have a similar range. Other projected new projectiles include a shell carrying 195 fuzed grenades, to be discharged over the target area, a chemical projectile for screening smoke, and a 'binary' projectile which carries two containers of non-toxic chemicals which, under the stress of firing, rupture and blend so as to form a nerve gas agent. The M 110 A2 system is expected to be in service by the early 1980s.

Employment	Belgium	Italy
	Germany (FRG)	Netherlands
	Iran	UK
	Israel	USA

8 inch SP howitzer M 110. Note exposed crew and hydraulic spades

Mortars

A mortar is basically a specialised form of howitzer, designed to fire solely at high-angle, i.e. 45–80°, and using graded charges of propellant to vary its trajectory. The piece, generally smooth bore, is supported at the muzzle end by a bipod and rests on a circular steel base plate which takes all the shock of discharge. Typical mortars have no buffer-recuperator mechanism, which enormously simplifies manufacture. Recoil is kept low by using relatively small propellant charges as compared with a howitzer, which enables the whole equipment to be light and, in the smaller natures, man-portable, but limits the range. The noise of discharge is almost inaudible in battle conditions, and the high angle of projection makes the mortar easy to deploy, particularly in jungle or in mountains. The steep angle of descent enables deep valleys, reverse slopes and jungle clearings to be searched.

The projectile is a finned bomb which is loaded and fired by simply dropping it down the muzzle by hand, the primer hitting a fixed firing pin at the bottom of the breech. (Alternatively, in some models as mentioned below, trigger firing is possible; 'controlled' as opposed to 'automatic' discharge.) The thin walled bomb has excellent blast fragmentation and is also a good smoke and chemical vehicle.

Laying is indirect in the ordinary way, a small dial-sight referring to an aiming post stuck in the ground a few yards off.

The mortar is a quick-reaction weapon. It is normally deployed well forward and fire-control can either be by voice from an observation post a short distance ahead, or by man-pack radio. The zone of dispersion is a broad ellipse or even near-circular and relatively large in terms of range as compared with a breech-loading rifled piece, but the rate of fire is high—20 rounds per minute from well trained crews—and as a simple, inexpensive device for bringing down a saturating area concentration within its limits of range the mortar is unequalled. Used in ones and twos close behind the attacking troops it is extremely useful for rapid close support on small targets in squad and platoon action. The great advantage of the mortar is its small size, mobility, cheapness and simplicity. Not surprisingly it is the favourite weapon of guerrillas, being extremely difficult to locate and counter-bombard. As will be seen in the ancillary section pp. 141–143 a specialised locating equipment has been developed for the location of enemy mortars.

The mortar is the chosen self-operated close support weapon of the infantry (who took to it whole heartedly in World War II, discarding the infantry howitzers and guns in the cannon-companies of the infantry regiments where these existed in its favour). Nevertheless the larger natures, say, from a calibre of 105 mm upwards, have true artillery characteristics regardless of the affiliation of the operating arm. The Israeli army has a 160 mm mortar, and 120 mm mortars can now range up to 9,000 m, or 5·4 miles. Except for the British, who have set their face against any but light infantry mortars, they are deployed by many armies as an essential complement to the other two members of the indirect fire artillery family—the guns and rockets.

The passion for sophistication has also overtaken mortar designers. Mortars are being made more accurate by improving their sights and ballistics. Mortar ammunition is becoming more and more expensive as lethality is increased and range and consistency improved by rifling, and plastic driving bands, which expand on discharge to seal the bore effectively; the bomb must, of course, be a loose fit so as to slide down the barrel in the act of loading. There is also a demand for the greater ranges provided by the rocket-assisted projectile principle. Hotchkiss-Brandt, for instance, have developed their *Projectile*

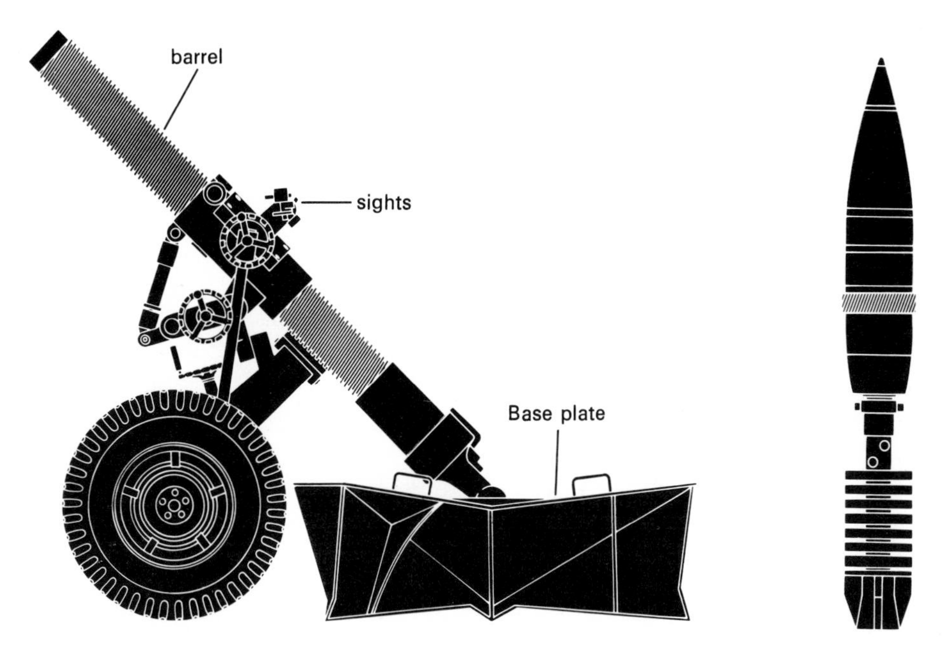

Figure VIII(a): Typical heavy mortar

Figure VIII(b): PEPA mortar round

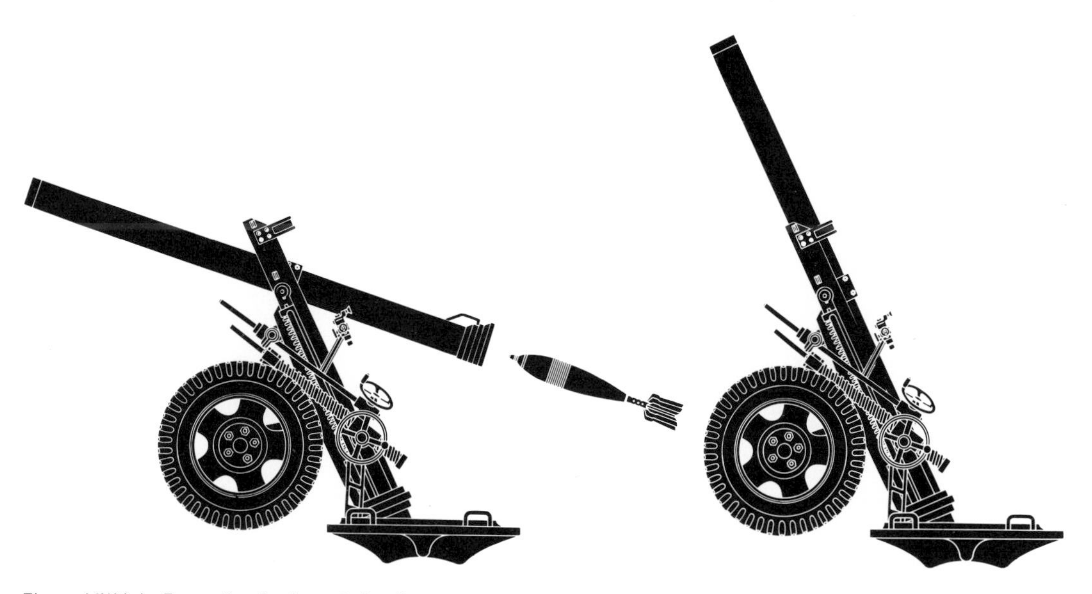

Figure VIII(c): Example of a breech-loading mortar

Empenné á Propulsion Additionelle (PEPA) to improve the range of their 120 mm mortar from 6,650 m to 9,000 m and the Soviet artillery has a huge breech-loader.

Here, of course, we have a characteristic design dilemma. The general utility of mortars —their cheapness, simplicity of operation, mobility, flexibility of deployment, high power and lethality—have made demands for bigger mortars, more accuracy, and more range so that cost and complication increase and indeed some of the rare, very large mortars that have been designed are really more like high-angle howitzers. This in turn generates a demand for sophisticated fire control procedures and target acquisition equipment. The opinion of artillerymen as a whole is to resist this trend and to maintain the distinction between the crude, simple mortar and sophisticated, rifled, breech-loading ordnance.

105 mm EC1A 105-L medium mortar

Spain

Calibre	105 mm
Barrel length	1,500 mm
Barrel weight	—
Baseplate weight	—
Bipod weight	—
Total weight in action	105 kg
Ammunition options	HE (9·2 kg)
	Smoke (9·2 kg)
	Practice (9·2 kg)
Range	7,050 m
Rate of fire	12 rds/min
Elevation limits	—
Traverse limits	—
Detachment	4

105 mm EC 1A in firing position

This mortar is of similar design to the ECIA 81-L mortar. The differences are in calibre, the strength of the various parts, and a differently shaped baseplate, which is round in this case.

The barrel is mounted on the baseplate by a ball and socket joint, and is supported by a tripod instead of the more usual bipod.

Due to the extra weight over the 81-L, two types of trailer have been designed for this weapon. The normal trailer of box-type construction is capable of carrying the complete weapon, its accessories and twelve rounds of ammunition. It weighs 239 kg without ammunition. A light cart is available for airborne operations; it weighs only 110 kg, and will carry the whole weapon, but no ammunition. Though it is normally transported on a trailer, this mortar can be carried by three men or by one mule.

The HE round has a damage radius of 150 m from its 1·75 kg charge of TNT. A smoke round is available.

Employment Spain

The 105 mm EC 1A on its trailer with 12 rounds of ammunition

107 mm mortar M 30

United States of America

Calibre	107 mm
Barrel length	1,520 mm
Barrel weight	71 kg
Baseplate	100 kg (M 24)
Total weight in action	295 kg
Ammunition options	HE (M 329) (11·2 kg)
	HE (M 3)
	Smoke WP
	Chemical
	Illuminating
Range	5,420 m
Rate of fire	20–25 rds/min
Elevation limits	+45° +85°
Traverse limits	16°
Detachment	5–6

The M 30 is a muzzle-loaded rifled weapon, introduced into service in 1951, which has two mountings. The difference lies in the basplate: M 24 is an inner baseplate of magnesium with an outer baseplate ring: M 24 A1 is a one-piece steel assembly and weighs 104 kg. The whole equipment weighs 295 kg, and breaks into six parts for transporting: barrel with integral tube cap and firing pin; baseplate; base ring; bridge; rotator; monopod support or standard. The monopod support is an important recognition feature.

Two types of HE ammunition are available: M 329 (and M 329-B1) ranges to 5,420 m with a muzzle velocity of 293 m/sec: M3 (and M 3A1) gives a muzzle velocity of only 257 m/sec and achieves a range of 4,600 m. The chemical round gives the same performance as M 3. This mortar is used in an SP configuration when mounted in a modified M 113 APC known as the M 106.

Employment		
	Austria	Netherlands
	Italy	Philippines
	Japan	USA
	South Korea	Zaire

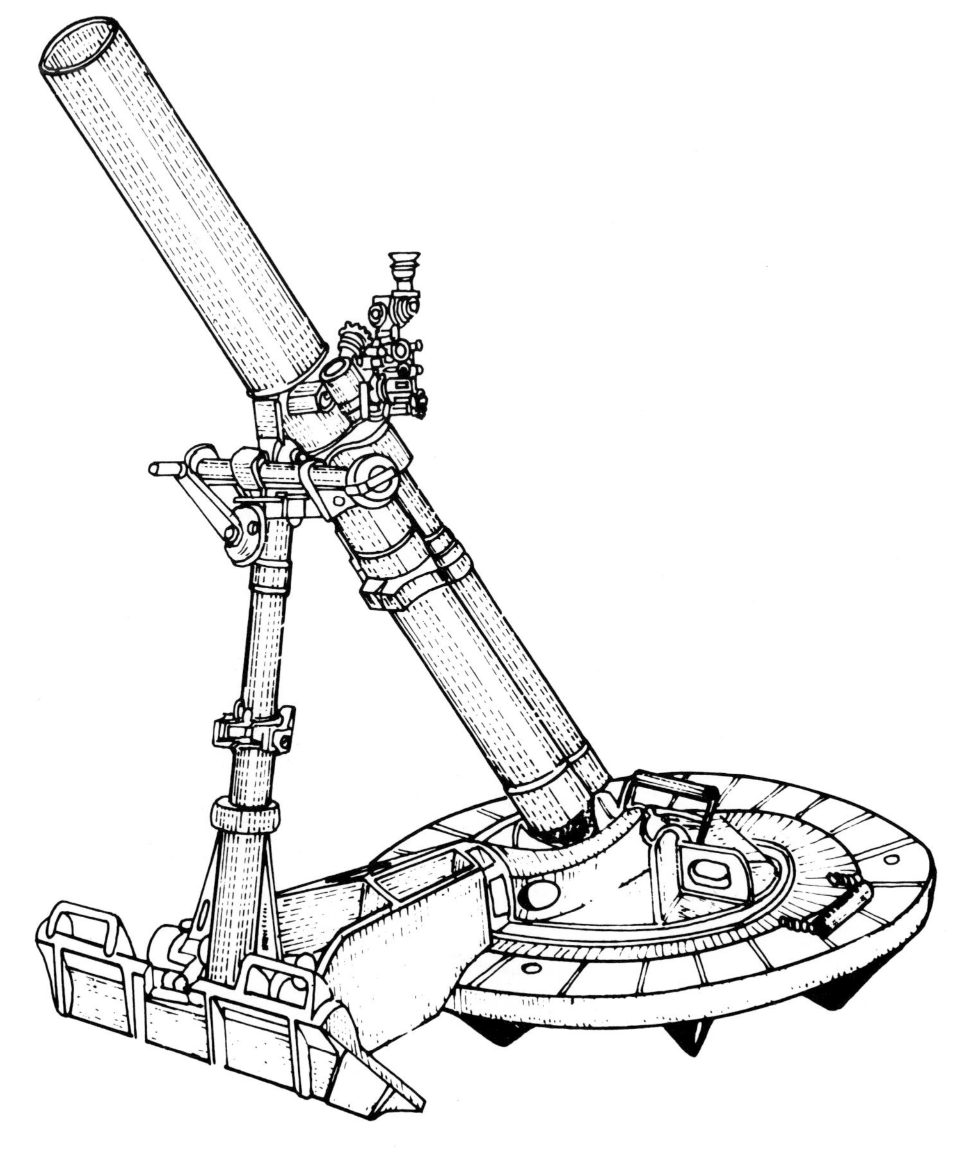

107 mm mortar M 30

107 mm heavy mortar M 1938 and M 107

Union of Soviet Socialist Republics

Calibre	107 mm
Barrel length	1,670 mm
Barrel weight	—
Bipod weight	—
Baseplate weight	—
Total weight in action	170 kg
Ammunition options	HE (9 kg)
	HE light (7·9 kg)
Range	6,300 m
Rate of fire	15 rds/min
Elevation limits	+45° +80°
Traverse limits	6°
Detachment	6

The 107 mm mortar of two marks, M 1938 and M 107, is a scaled down version of the Soviet 120 mm mortar, reduced in weight and size to suit its normal role, as the regimental mortar of mountain units. The mortar is usually transported by pack animals in difficult terrain, but is towed on a two-wheeled carriage over normal roads. For animal-packing, the weapon breaks down into three loads; barrel, bipod and baseplate.

The M 107 is the improved model of this mortar and has replaced the M 1938 in the Soviet Army, although the earlier version is still to be found in the non-Soviet Warsaw Pact countries and in South-East Asia.

Employment China, Ethiopia, Germany (GDR), India, Jordan, North Korea, Poland, USSR, Vietnam

The 1938 model 107 mm mortar deployed

The M 107 mortar in travelling position

4·2 inch mortar Mk 3

United Kingdom

Calibre	107 mm
Barrel length	1,980 mm
Barrel weight	41·3 kg
Bipod weight	20·8 kg
Baseplate weight	84 kg
Total weight in action	550 kg (including carriage)
Ammunition option	HE (9·1 kg), Smoke (10·2 kg)
Range	3,750 m
Rate of fire	12 rds/min
Elevation limits	45° – 80°
Traverse limits	7° – 21°
Detachment	6

The British 4·2 inch mortar first saw action at El Alamein and proved to be a most effective weapon in World War II. It is still in service in five armies. It consists of a steel tube with a cap and breech piece screwed onto it, a bipod, a cradle with elevating and traversing gears, sights and a base plate. It has a two charge ammunition system. There are six men in the detachment with two ¾ ton type towing vehicles: one towing the mortar and the other towing a 10 cwt two wheeled ammunition trailer, which holds up to 44 bombs.

Employment Ethiopia, Laos, Malaysia, Nepal, Turkey

The 4·2 inch mortar in action: note square spade attached to base plate

Austria

Calibre	120 mm
Barrel length	1,500 mm
Barrel weight	—
Bipod weight	—
Baseplate weight	—
Total weight in action	296 kg
Ammunition options	HE (15·8 kg)
Range	5,700 m
Rate of fire	15 rds/min
Elevation limits	+45° +80°
Traverse limits	6°
Detachment	6

This Austrian weapon of unknown ancestry consists of a conventional baseplate, bipod and buffer design. It is removed from its heavy two wheel trailer for firing. Unlike many other designs, the weapon does not form part of the trailer carriage, to be towed by its muzzle, but rather sits on a complete trailer.

The barrel is mounted by ball and socket to the baseplate to allow 360° traverse. The baseplate is dish-shaped with spade like extensions beneath; it has two integral lifting handles.

Employment Austria

---

120 mm standard Tampella mortar (M 65)

Finland

Calibre	120 mm
Barrel length	1,940 mm
Barrel weight	84 kg
Bipod weight	69 kg
Baseplate weight	72 kg
Total weight in action	225 kg
Ammunition options	HE (M 58F) (12·9 kg) HE (ST) Smoke Illuminating Practice
Range	8,300 m
Rate of fire	5–10 rds/min
Elevation limits	+37° +87°
Traverse limits	360°
Detachment	3–4

120 mm Tampella mortar mounted on the M 3 half-track

The 120 mm standard Tampella is a smooth-bore mortar made to high tolerances and is designed for simplicity in both operation and maintenance. Though normally towed, it can be carried by mule, or man-handled over very short distances.

The barrel is made of high-tensile steel alloy and has a close tolerance bore. It is internally threaded at the breech end to take the breech block which contains the firing mechanism. The firing mechanism is of the spring-loaded striker type actuated by a lanyard and has a safety lever. The bipod incorporates the recoil buffer with the barrel collar, elevation, azimuth and correction gears. Leg spread is limited by an adjustable chain. Welded sheet steel is used for the 960 mm diameter baseplate, in the centre of which is a socket to take the breech-piece ball end. The mortar can be traversed 360° without moving the baseplate. The sight unit consists of an elevation and azimuth mechanism, one range and two cross-levelling bubbles and a collimator. Quick releases for large adjustments are fitted and the mils-graduated dials can be freed for rapid and large adjustments. A quick release dovetail mount on the bipod facilitates removal of the sight, though this is not necessary when firing. A steel box chassis two-wheel trailer is normally used to move the weapon. It weighs 129 kg.

Four types of ammunition are available, though any 120 mm mortar ammunition can be used. The M 58F round is a HE bomb containing 2·3 kg of TNT and utilises up to eight secondary charges. Three types of smoke round of the same designation are available (FM, PWP and WP) also illuminating and practice rounds. With the Super Tampella (ST) bomb, a range of 8,300 m is possible.

The Israeli Army until recently relied almost entirely on mortars for close support artillery, and it is this weapon—made under licence by Soltam—which is used. It is carried in, and often fired from, the M 3 half-track and is deployed in regiments of 12 mortars with each infantry brigade.

Employment Finland
Germany (FRG)
Israel

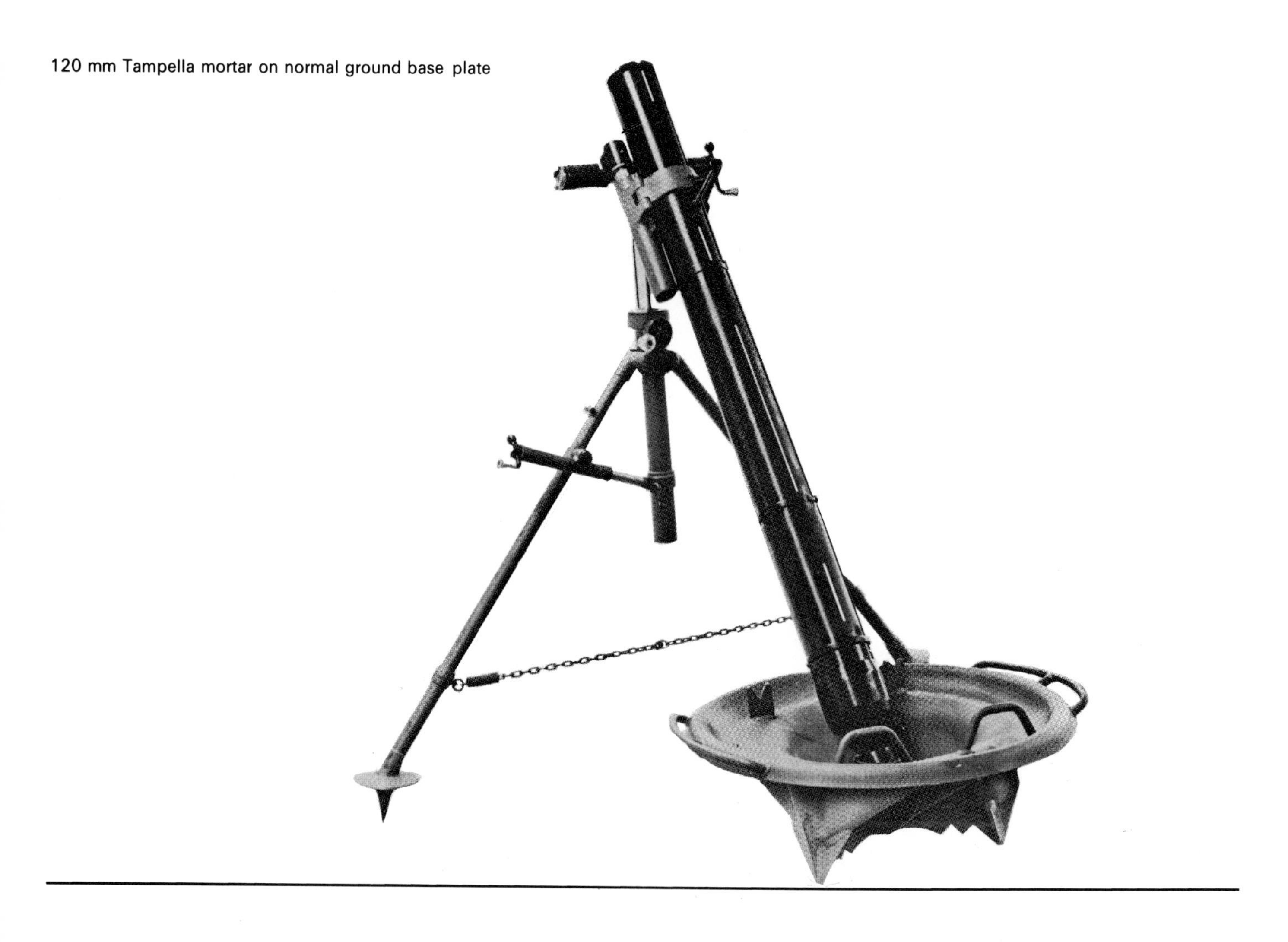
120 mm Tampella mortar on normal ground base plate

120 mm light Tampella mortar

Finland

Calibre	120 mm
Barrel length	1,726 mm
Barrel weight	43 kg
Bipod weight	31·5 kg
Baseplate weight	62 kg
Total weight in action	136 kg
Ammunition options	HE (M 58F) (12·9 kg)
	HE (M 58FF) (13 kg)
	Smoke
	Illuminating
	Practice
Range	6,250 m
Rate of fire	5–10 rds/min
Elevation limits	+45° +70°
Traverse limits	—
Detachment	3

The 120 mm light Tampella mortar is a smooth-bore weapon with a retractable firing pin. The barrel is made of steel alloy, and is externally threaded at the breech to take the breech block containing the firing pin. A ball extension on this block makes the joint with a central socket in a circular welded baseplate and enables the mortar to be fully traversed without re-bedding. The bipod is of similar design to all the Tampella mounts, and incorporates a recoil buffer, and dustproof gearing for elevation, azimuth and fine adjustment. The sight unit is the same as that of the 120 mm standard, and of the 81 mm Tampella mortar, and is covered in the description of the former.

The optional carriage is of tubular construction and weighs 130 kg. It can be towed at speeds up to 40 km/hr. The weapon, all its accessories and six rounds of ammunition can be carried. The mortar can be man-packed by three men or one mule.

A special range of ammunition designated M 58FF has been produced for this mortar, though it will also fire the M 58F ammunition. The HE bomb is filled with 2·8 kg of TNT. Other rounds include three types of smoke, and illuminating and practice bombs. All rounds have the same ballistic properties, so only one range table is necessary.

This mortar, like the Standard Tampella M 65, is made in Israel.

Employment Finland
Israel

(illustrated overleaf)

120 mm light Tampella mortar deployed

In travelling position

120 mm MO-120-AM 50 heavy mortar

France

Calibre	120 mm
Barrel length	1,746 mm
Barrel weight	76 kg (with towing ring)
Bipod weight	—
Baseplate weight	80 kg
Total weight in action	402 kg (242 kg without undercarriage)
Ammunition options	HE (PEPA/LP) (13·4 kg) HE (M 44) (13 kg) Coloured smoke (13 kg) Illuminating (Mk 62 ED) (13·6 kg) Smoke WP (Mk 62) (13 kg) Practice
Range	9,000 m (with PEPA ammunition)
Rate of fire	8–12 rds/min
Elevation limits	+45° +80°
Traverse limits	17°
Detachment	4

This is a smooth-bore mortar capable of both automatic and controlled fire. A bipod is provided for greater stability during sustained fire, but it can also be fired from its road wheels which are equipped with a special locking device to stop the weapon moving when it is fired without its bipod. In its travelling position a screw-on ring mount is attached to the muzzle end of the tube, and the whole unit is towed by a suitable vehicle. The AM 50 fires the same family of ammunition as the other Hotchkiss-Brandt 120 mm mortars, including the rocket assisted PEPA round which produces the maximum range of 9,000 m.

Employment France

Note the triangular base plate and the use of the wheeled carriage to form a bipod support

Calibre	120 mm
Barrel length	1,632 mm
Barrel weight	34 kg
Bipod weight	25 kg
Baseplate weight	33 kg
Total weight in action	92 kg
Ammunition options	HE (PEPA/LP) (13·4 kg) HE (M 44) (13 kg) Coloured smoke Illuminating (M 62) (13·6 kg) Smoke WP Practice
Range	5,500 m 6,550 m (PEPA)
Rate of fire	8–15 rds/min
Elevation limits	+40° +85°
Traverse limits	17°
Detachment	3

The MO-120-60 is the lightest of a range of 120 mm mortars manufactured by Hotchkiss-Brandt. Its total weight of 94 kg permits it to be broken down into a three-man load. It is also APC mounted in AMX VCL. It is conventionally designed smooth bore weapon with most refinements eliminated in order to reduce as much weight as possible. A unique buffer absorbs some of the shock of firing, and has probably been evolved to reduce the stress on the very light construction. The barrel weight of 34 kg indicates that it is thin-walled and the maximum charge (4) suggest that this light barrel structure imposes a limitation on range. Nevertheless, a wide choice of ammunition is available for this weapon. The M 44 HE bomb is the primary round; others include coloured HE, smoke, illuminating and training. The bombs weigh between 13·8 kg and 13 kg. The rocket assisted PEPA round is available, which ranges to 6,550 m.

Employment Argentina
France

Hotchkiss-Brandt 120 mm (MO-120-60) mortar

120 mm MO-120-M 65 heavy mortar

France

Calibre	120 mm
Barrel length	1,640 mm
Barrel weight	44 kg
Bipod weight	24 kg
Baseplate weight	26 kg
Total weight in action	104 kg
Ammunition options	HE (PEPA/LP) (13·4 kg) HE (M 44) (13 kg) Coloured HE Illuminating (Mk 62) (13·6 kg) Smoke WP Practice
Range	9,000 m
Rate of fire	8–12 rds/min
Elevation limits	+40° +85°
Traverse limits	17°
Detachment	2–3

The MO-120-M 65 is an attempt to combine some of the lightness of the MO-120-60 with the fire power of the MO-120-AM 50. Normal transportation is achieved by putting the weapon on a light trailer which can be towed by a vehicle or pulled by two men. However, due to the relatively light weight, it is possible to manpack the weapon in three roughly equal loads, using special harness.

This is a smooth-bore weapon fired from a tripod and baseplate and capable of either automatic or controlled operation. In addition, the breech is equipped with a safety device to prevent accidental percussion when, for example, the weapon is being unloaded. The weapon is capable of firing any 120 mm ammunition, including the PEPA/LP (projectile empenné à propulsion additionelle/longue portée) HE round with additional propulsion.

Employment France

Hotchkiss-Brandt 120 mm (M 65) mortar on its light trailer

120 mm MO-120-RT-61 heavy rifled mortar

France

Calibre	120 mm
Barrel length	2,080 mm with breech
Barrel weight	114 kg
Bipod weight	—
Baseplate weight	190 kg
Total weight in action	582 kg
Ammunition options	HE (PRPA) (15·6 kg) HE (PR 14) (15·7 kg) Preclair illuminating (15·7 kg)
Range	13,000 m with PRPA, 8,350 m with PR14
Rate of fire	6–10 rds/min
Elevation limits	+30° +85°
Traverse limits	14°
Detachment	3

The Hotchkiss-Brandt 120 mm rifled mortar is one of the largest mortars available today. It is mounted and towed on a specially designed pair of road wheels from which it is also fired. Stabilisation is achieved by the baseplate, which is effective both on hard and marshy ground. The manufacturers claim that the weapon can be brought into action in 1½ minutes and out again in two minutes.

The mounting, consisting of the cradle and carriage, is of advanced design. The traversing gear and cross levelling gear are similar to those found in the Hotchkiss-Brandt standard 120 mm mortar, but the elevating system is unique, in that quick, rough elevation is obtained by varying the position of the axis of the cradle relative to the axis of the wheels, and this having been done, fine adjustment is obtained by varying the position of the cradle, in relation to the barrel.

The weapon is capable of either automatic or controlled fire and has its own range of ammunition. The PR 14 HE bomb is comparable in its effect to the standard 105 mm artillery shell. The bomb weighs 15·7 kg and has a maximum range of 8,350 m. In addition, the PRPA (Projectile Rayé a Propulsion Additionelle), which weighs 15.6 kg, has a maximum range of 13,000 m due to the incorporation of a rocket motor which is initiated during flight. The Preclair illuminating round has a range of 8,000 m, and candle power of 1,050,000 for one minute or 1,600,000 for 40 seconds. Firing of conventional ammunition requires the use of range conversion tables since the other Hotchkiss-Brandt bombs are calibrated for different barrel lengths. The barrel of this mortar is rifled from breech to muzzle with 40 grooves with a uniform twist of 10·5°. The screw-on breech piece has a spherical ball butt flattened on one side which is locked into a socket in the baseplate. The firing mechanism is completely watertight and can be used submerged.

Employment France

120 mm MO-120-RT-61 mortar: note the simplicity of design: the mortar is towed by its muzzle

120 mm ECIA 120-SL and 120-L heavy mortar

Spain

	120 SL	120 L
Calibre	120 mm	120 mm
Barrel length	1,600 mm	1,600 mm
Barrel weight	—	—
Baseplate weight	—	—
Bipod weight	—	—
Total weight in action	123 kg	213 kg
Ammunition options	HE (N Type) (16·7 kg) HE (L Type) (13·2 kg) Smoke (N Type) (16·7 kg) Smoke (L Type) (13·2 kg) Practice (N & L Types)	HE (N Type) (16·7 kg) HE (L Type) (13·2 kg) Smoke (N Type) (16·7 kg) Smoke (L Type) (13·2 kg) Practice (N & L Types)
Range	5,940 (L type)	6,600 (L type)
Rate of fire	12 rds/min	10 rds/min
Elevation limits	—	+45° +85½°
Traverse limits	—	—
Detachment	4	4

There are two versions of the EC1A 120 mm mortar.

The ECIA 120-SL has a similar configuration to other ECIA mortars such as the 81-L and 105-L; it has the same tripod support. A feature of this mortar, not apparent on the other members of the ECIA family, is an adjustable chain for limiting the leg spread of the tripod. Although normal transport is by trailer, the manufacturer claims that it is possible for the weapon to be manpacked by three men.

Two types of ammunition designated N and L are available for this weapon. The N type bombs are heavier and longer, and carry a bigger charge (3·2 kg of TNT or smoke composition), than the L bomb (2·3 kg).

In firing position

The range of the N type is 900 m less than the L type for both weapons. Packaging of ammunition is the same in each case. Two rounds are packed in individual containers in a wooden case.

The ECIA 120-L is a longer range version of the 120-SL. The basic configuration is the same, although it has a larger baseplate and a heavier barrel. It fires the same ammunition as the 120–SL and is transported on a 146 kg trailer. It is too heavy for man-packing. The other difference between the two mortars is that the 120-L does not have the adjustable chains to limit tripod leg spread.

Employment Spain

Mounted on its trolley

120 mm mortar M 41C

Sweden

Calibre	120 mm
Barrel length	2,000 mm
Barrel weight	—
Bipod weight	66 kg
Baseplate weight	—
Total weight in action	285 kg
Ammunition options	HE (M 58F) (12·9 kg) Smoke Illuminating Practice
Range	6,400 m
Rate of fire	12 rds/min
Elevation limits	+45° +80°
Traverse limits	360°
Detachment	4

This is a Second World War weapon originally designed and produced in Finland as the Tampella M 1940. In 1960, the original base plate was replaced by the Hotchkiss-Brandt M 52, locally manufactured under licence from France.

The bipod is the same as that fitted to the Tampella M 65. It incorporates the twin recoil buffer and laying gear. A cross bar in front of the barrel collar carries the sight at its left end and traversing hand wheel at the right. Leg spread is limited by an adjustable chain. The carrying trailer is detached for firing.

The barrel carries lifting handles near its base and is mounted into the centre of the baseplate. The whole equipment is traversed about the baseplate mounting to give 360° traverse.

The M 41C is 65 kg heavier than the later M 65, with which it has shared a common sight since 1972. The M 58F range of ammunition can be fired but it is not known if the longer range super-Tampella round is accepted. The ammunition types include three different smoke rounds.

Employment Eire
Sweden

120 mm mortar MW 41

Switzerland

Calibre	120 mm
Barrel length	—
Barrel weight	—
Bipod weight	—
Baseplate weight	—
Total weight in action	262 kg
Ammunition options	HE (13 kg)
Range	5,000 m
Rate of fire	10 rds/min
Elevation limits	—
Traverse limits	—
Detachment	6

The MW 41 is an elderly and heavy weapon characterized by the convex baseplate with four hinged carrying handles. The barrel is fixed to the baseplate by a conventional ball and socket joint. The bipod, mounted to the barrel with a recoil buffer, is also of a conventional pattern. It allows limited traverse, and elevation from about 45° to 80°.

Reconnaisance battalions and armoured regiments of mechanised brigades are equipped with this mortar.

Employment Switzerland

120 mm heavy mortars M 1938 and M 1943

Union of Soviet Socialist Republics

Calibre	120 mm
Barrel length	1,850 mm
Barrel weight	—
Bipod weight	—
Baseplate weight	—
Total weight in action	500 kg
Ammunition options	HE (15·4 kg) Smoke (16 kg) Incendiary (16·7 kg)
Range	5,700 m
Rate of fire	12–15 rds/min
Elevation limits	+45° + 80°
Traverse limits	8°
Detachment	5–6

These two weapons are almost identical. Detailed modifications have been made, but do not affect the MP-41/MP-42 collimating sight. The 120 mm mortar M 1938 was the standard Soviet infantry mortar until it was replaced recently by the 120 mm M 1943. A unique design when originated, the mortar can be quickly broken down into three parts, barrel, bipod and baseplate, for movement over short distances. For normal travel the whole weapon folds together and is towed on a two-wheeled tubular carriage or, if necessary, can be animal-packed in its three parts. The mortar can be either drop or trigger-fired by lanyard, as required, and an anti-double-loading device is provided for attachment to the muzzle.

Although the 120 mm mortar M 1943 has virtually replaced the M 1938 as the regimental mortar of the Soviet infantry both are found in the other countries listed. The M 1943 is manufactured in communist China as the 120 mm mortar type 55, and exported to Pakistan and Tanzania.

Whilst the ballistic and performance details and methods of handling are the same for both models, the outward appearance differs slightly: the shock absorber cylinders of the M 1943 are much longer than those of the M 1938, and in the later model the elevating and traversing gear is rather more sophisticated.

The outer casing of the HE ammunition can be made of either wrought or cast iron, the latter being more effective against personnel but being slightly reduced in maximum range.

The Austrians refer to this mortar as M 60.

120 mm heavy mortar M 1943

Employment

Afghanistan
Algeria
Albania
Austria
Bulgaria
China
Congo
Czechoslovakia
Egypt
Germany (GDR)
Iraq
Lebanon
Morocco
North Korea
Pakistan
Rumania
Spyria
Tanzania
USSR
Vietnam
Yemen
Yugoslavia

---

120 mm UBM 52 heavy mortar

Yugoslavia

Calibre	120 mm
Barrel length	1,290 mm
Barrel weight	—
Bipod weight	—
Baseplate weight	—
Total weight in action	400 kg
Ammunition options	Heavy HE (15·9 kg) Light HE (12·25 kg) Smoke Illuminating (12·2 kg)
Range	6,000 m (with the lighter shell)
Rate of fire	25 rds/min
Elevation limits	+45° +85°
Traverse limits	60 at 45° elevation
Detachment	5

The Yugoslav 120 mm mortar UBM 52 can fire the same ammunition as the Soviet 120 mm mortars but is an adoption of a Western weapon, the MO-120-AM 50 made under licence. The UBM 52 was designed as a

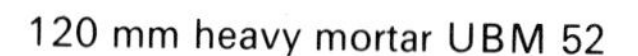

120 mm heavy mortar UBM 52

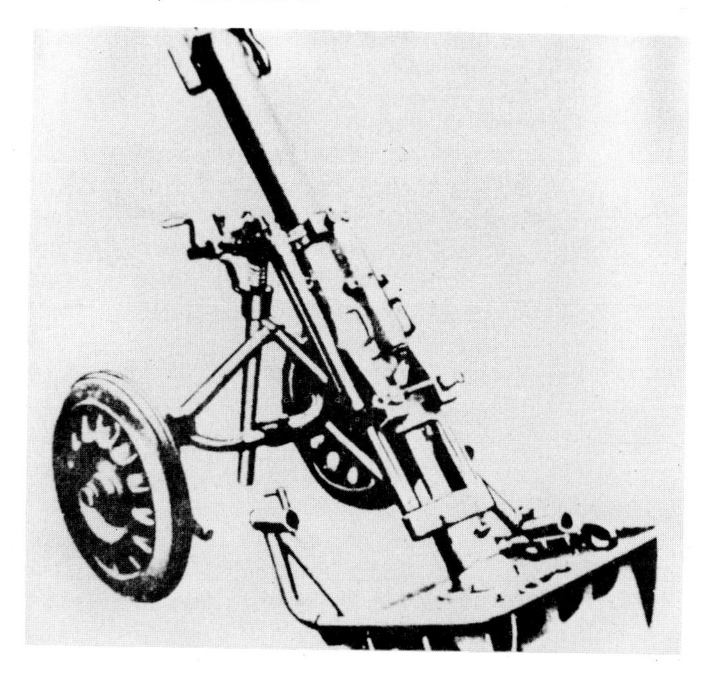

heavy mobile weapon for use both in normal operations and in mountainous terrain. It is thus semi-permanently mounted on its own two-wheeled carriage for firing, but can be taken apart for pack transport by five mules. It also possesses a muzzle attachment to permit towing by either two mules or a vehicle. The cleverly designed spiked rectangular baseplate and hydro-elastic recoil system allow the mortar to be halted and fired almost immediately, no ground preparation for the baseplate being necessary. In spite of its shorter barrel, the UBM 52 can outshoot the Soviet M 1943: using its light HE ammunition, it has an extra 300 m range and a far higher rate of fire.

Employment Burma
Yugoslavia

---

160 mm Tampella heavy mortar

Finland

Calibre	160 mm
Barrel length	3,066 mm
Barrel weight	—
Bipod weight	—
Baseplate weight	250 kg
Total weight in action	1,700 kg
Ammunition options	HE (40 kg) Smoke
Range	9,600 m
Rate of fire	5–8 rds/min
Elevation limits	+43° +70°
Traverse limits	360°
Detachment	7

This is a smooth-bore weapon which incorporates a number of novel features. It is mounted on a two-wheeled carriage, but the baseplate is removed and loaded in the towing vehicle for travelling. In action, the wheels are swung inwards so as to be at a tangent to the baseplate centre. This endows the weapon which is too heavy for the detachment to lift, with 360° traverse. The movement of the carriage wheels is used for both coarse and fine azimuth laying: for fine laying a hand wheel and reduction gear is fitted to the left hand wheel.

The barrel, breech piece and recoil buffer are of conventional design.

The HE round contains 5 kg of explosive (TNT) and is fitted with an impact fuse. A primary and nine secondary charges are used. The barrel is lowered for loading: a semi-automatic lowering device and a counter-balance mechanism enable the detachment to produce a high rate of fire for the size of the equipment.

The weapon is made under licence in Israel by Soltam Ltd. Many of the Israeli 160 mm mortars are mounted on a modified Sherman tank chassis. A ramp is used to present ammunition to the barrel for loading from its on-board stowage.

Employment Finland
Israel

160 mm Tampella heavy mortar

240 mm mortar M 53

Union of Soviet Socialist Republics

Calibre	240 mm
Barrel length	5,340 mm
Barrel weight	—
Bipod weight	—
Baseplate weight	—
Total weight in action	3,610 kg
Ammunition options	Nuclear (130 kg) HE (100 kg)
Range	10,000 m
Rate of fire	1 rd/min
Elevation limits	+45° +65°
Traverse limits	17°
Detachment	8

This is the largest mortar in service today, but is believed to be obsolete in the Soviet Army. It is a breech loaded weapon made up of baseplate, two-wheeled carriage, barrel and cradle. The baseplate is a conventional disc with a diameter of 2,130 mm. The wheeled carriage supports the cradle and barrel, serving in place of the bipod of smaller mortars. Elevation is adjusted by gearing at the attachment of cradle to carriage. The barrel is mounted at its centre of gravity in trunnions set in the top of the cradle: this allows the barrel to be swung to the horizontal for loading.

The general arrangement allows reasonable elevation, but limits traverse. Large switches present a problem because of the size of the weapon. The fin-stabilised bomb is loaded through a vertically sliding breech block. Lifting the bomb to the breech is a four-man task using a giant Stilson clamp with two long handles. The breech stands over five feet from the ground in the loading position. A small trolley is used to move rounds about the baseplate position.

240 mm mortar M 53 with barrel depressed for loading

Little is known about the nuclear round, which is probably comparable with the American 8 inch howitzer shell. The short range of the mortar must present a problem in deployment in the nuclear role.

Employment		
	Algeria	India
	Bulgaria	Iraq
	China	Rumania
	Hungary	USSR

In travelling position: note umbrella-shaped base plate on left of picture. The mortar is towed by the barrel

VI. Artillery Rockets

This section deals with a wide variety of rockets, both guided and free flying, with ranges from a few thousand metres to hundreds of kilometres, and warheads from two or three kilogrammes to megaton nuclear devices. The criteria for inclusion in this chapter is that the weapon should be truly mobile and intended to influence the immediate battle rather than targeted for strategic effects. The Attila rockets are hardly in the same class as the Scaleboard or Shaddock: nor are the multiple launchers strictly comparable in purpose with single guided weapons, except that there is an interesting link weapon in the Japanese 30 rocket. All types of rocket are included together for convenience.

Some missiles have guidance systems, and most have fins, but these are added to the two basic components—rocket motor and warhead. For small missiles, especially those for which only one type of warhead is available the two parts will be pre-assembled and delivered as a complete unit. Larger missiles in the Honest John/Lance class will almost certainly be delivered to the users as separate components. The assembly area can be away from the launch site for security, and the time of assembling is also used for checking out the various missile systems.

In the case of liquid fuel motors, fuelling is usually left until the last moment. It is a hazardous and complex task requiring bulky equipment. Warhead mating and fuelling will probably be at the launch site (an exception is Lance whose liquid fuel is pre-packed and ready for use) because of the difficulty of moving the missile after fuelling.

Many rockets have wings, or more accurately fins. These are not aerodynamic and do not generate lift, but are control surfaces. Unguided rockets use them to give stability in flight: they are slightly offset from the axis of the rocket and sustain, in flight, the spin imparted at launch. Guided missiles employ moveable control surfaces which effect changes in the missile's course.

Either solid or liquid fuels are used to power the rocket motor, each having its own advantages and disadvantages. Liquid fuel is dangerous to handle, difficult to store and requires a sophisticated control and motor design, yet it has a greater energy output for its weight, thrust can be varied and the motor stopped and restarted. The solid fuels have the opposite qualities. In general, the smaller and cheaper rockets are solid fuelled.

A free flight rocket is inherently less accurate than a projectile fired from a gun. The sources of inaccuracy can be minimised in design and manufacture, although this increases the cost of the weapon, to a point where it is economically sensible to adopt a guidance system. As a result the tactical handling of rockets is designed to take advantage of the weapon's characteristics.

What are the sources of inaccuracy?

First, the behaviour of the rocket motor. Unless some of the advantages in warhead design (which will be explained later) are compromised, or the launch rail is excessively long, the motor will continue to burn while the missile is in free flight. This means that the missile is subjected to the influence of crosswind and low-level air eddies while the centre of gravity and thrust is changing because the propellant is being consumed. This effect is magnified by any misalignment or thrust due to manufacturing tolerances.

Second, the missile which consists of rocket motor and warhead is much larger than a shell, and therefore more susceptible to the effects of crosswinds. Stabilisation in flight normally depends on fins which increase the area against which the wind operates, and are especially productive of errors to the intended trajectory just after launch when the rocket is still accelerating.

Third, the burning time of the fuel is many times greater than that of a gun's propellant, so that small variations in the rate of burning have a more noticeable effect on the maximum velocity of the missile. This effect is exaggerated by the head or tail component of the wind: the wind acts against a shell only after it has achieved its maximum muzzle velocity in the protection of barrel, while the wind affects the missile's acceleration (and maximum velocity) as well as its unpowered flight.

Despite the relatively low accuracy of free flight rockets, they do have real advantages over the gun as a weapon delivery means. The rocket itself is not a cheap device, but its launcher is relatively simple compared with a gun. The production techniques of gun and shell require great industrial expertise in the construction of shell cases, barrels and stress bearing carriages. At the same time, they require sophisticated light industry for a host of ancillaries, from sights to hydraulics. The rocket requires only skilled light industry. As a result, rockets have been especially popular with the lesser industrial nations who have sought a measure of independence in their military needs: Brazil, Spain and Switzerland come to mind.

They have advantages too for the more sophisticated armies, which have them in service as well as guns. The Soviet Army especially has long seen their utility, and new Western weapons (LARS and RAP 14) will reflect the belief in a judicious mixture of the two weapon systems. The simple, unstressed rocket launcher can be assembled in banks, and once loaded, a large number of rockets can be ripple fired in a salvo lasting a few seconds. The shock effect at the target is great; only the mortar approaches it in this respect. No gun can achieve the same rate of fire, even the sophisticated Swedish L 50.

The rocket is less subject to limitations of calibre, is subject to smaller firing forces and therefore is less limiting to the warhead designer. As a result early nuclear weapons for the battlefield were rocket delivered. Multiple warheads and guided sub-munitions are more easily fitted to rockets. The motor shell provides a useful vehicle for controlled flight. Carrier warheads for smoke, incendiary, propaganda and chemical agents have a better payload than a shell of equivalent size.

The rocket's lack of accuracy can even be turned to advantage. The weight of fire which the multiple launcher produces allows the instantaneous saturation of a large area. This is especially effective with smoke, incendiary or chemical agent warheads, or against large or poorly defined targets. The effect of one salvo from a battery of LARS is equivalent to one round per gun from 10 battalions of guns.

There is a requirement to impose attrition on massed armoured attacks as early as possible in order to relieve pressure on direct fire weapons. The problem is that armour can only be defeated by a direct hit. Neither gun nor rocket can guarantee the accuracy required without resorting to some form of terminal guidance. The rocket's saturation capability, coupled to the freedom it allows the warhead designer provides an elegant solution. A multiple warhead consisting of small anti-armour devices not only increases the degree of saturation, and thus increases the chance of a hit, but attacks armour at its most vulnerable points—the roof and belly. Two types of sub-warhead are being worked on: the hollow charge bomblet or high energy solid shot attacking the roof, and perhaps more practical, the shaped charge minelet attacking the belly. This last has the additional advantage that it remains active after the critical moment of delivery. This concept is most advanced in the Pandora and Medusa warheads for LARS.

What has been said so far refers to small rockets which are directly comparable in range and warhead performance with guns. Post-war development has produced the giant 'single shot' weapons. These have more than filled the position in the order of battle once occupied by the super heavy artillery. They extend the ability of the army commander to influence the battle from his own resources beyond the range of conventional artillery. SCUD B, SCALEBOARD and Pershing are mobile weapons capable of prompt displacement after firing and originally designed to engage targets on the battlefield and in the rear areas, but their long range—to be increased in later marks of Pershing—has led to their inclusion in the category of 'theatre' weapons. They are therefore strictly speaking outside the scope of this book, but are included as they have this dual role.

The earliest 'battlefield missile' was a direct product of the World War II German work on the V2 type weapon. Guidance was achieved by radio signal. The process required the monitoring of the rocket's flight path by radar which increased the size of the handling unit, and was subject both to enemy jamming and deliberate misdirection. The latest types have inertial guidance systems which, being pre-programmed and independent of external signals, overcome both the earlier shortcomings. Today's battlefield missile is launched by a small section with few vehicles and is much more battle-worthy as a result.

Samurai

Italy

Calibre 76 mm
Rocket length 1,200 mm
Rocket weight 10·6 kg
Fuel Liquid
Stages One
Guidance Nil
Range 18,000 m
Warhead HE (3 kg)
Tubes/Rails —
Traverse limits —
Reload time —
Detachment —

The liquid propellant used in this inexpensive small calibre rocket is self pressurising and burns for 300 seconds. Maximum speed is 1,100 m/sec. The weapon is designed as a surface-to-surface, or air-to-surface weapon.

The cruciform fins at the rear have a 200 mm span.

Employment Trials only, Italy

80 mm 108-R rocket

Brazil

Calibre 80 mm
Rocket length —
Rocket weight —
Fuel Solid
Stages One
Guidance Nil
Range 12,000 m
Warhead HE
Tubes/Rails 16 tubes
Traverse limits —
Reload time —
Detachment

Very little is known about this small Brazilian rocket. Its calibre is in the order of 80 mm and it is fired from a multiple launcher.

The 16 tubes are arranged in three rows of five, six and five, and are mounted on a Land Rover type vehicle. It has no fins but is stabilised in flight by spin imparted by angled main motor vents.

Employment Brazil

81 mm Dira rocket

Switzerland

Calibre 81 mm
Rocket length 1,300 mm
Rocket weight 16 kg
Fuel —
Stages —
Guidance Nil
Range 8,700 m
Warhead —
Tubes/Rails 30 tubes
Traverse limits 360°
Elevation limits —
Reload time —
Detachment —

The DIRA rocket is spin and fin stabilised. It fires from an Oerlikon multiple launcher which is a 30 tube configuration and can be mounted on the M 113 APC. The tubes are arranged in two bundles of 15. It is also employed in a naval role and is suitable for deployment with airborne forces in a towed version.

Missile velocity is 490 m/sec and a warhead payload of 7 kg can be carried.

Employment Switzerland

81 mm Dira rocket

R-6-B-2 rocket

Spain

Calibre	108 mm
Rocket length	935 mm
Rocket weight	19·4 kg
Fuel	Solid
Stages	One
Guidance	Nil
Range	10,000 m
Warhead options	HE (8·8 kg) Smoke Incendiary
Tubes/Rails	32 cages
Traverse limits	45°
Reload time	8 mins
Detachment	——

The R-6-B-2 is a member of a family of Spanish surface-to-surface free flight rockets. Characteristics which are common to the whole family include the use of IMI motors, spin stabilisation, and a claimed delivery error of one per cent of range. The launching cages incorporate helical rails which impact spin. The motors are initiated electrically, and these R-6-B-2 rockets are launched at one second intervals.

The 32 launching cages are arranged in four rows of eight on a pedestal mounting standing on a flat platform one-ton truck. A stabilising jack is fitted to the vehicle.

Six launchers make up a battery.

Employment Spain

108 mm R-6-B-2 launcher elevated

110 mm Lars

Federal Republic of Germany

Calibre	110 mm
Rocket length	2,260 mm
Rocket weight	35 kg
Fuel	Solid
Stages	One
Guidance	Nil
Range	14,000 m
Warhead	HE (17·2 kg) Smoke Incendiary Anti-vehicle Anti-armour minelets
Tubes/Rails	2 boxes of 18 tubes
Traverse limits	210°
Reload time	15 mins
Detachment	5

Lars is a vehicle mounted rocket system designed to produce a high concentration of fire very rapidly.

The long rocket motor produces a velocity of 635 m/sec resulting in a flat trajectory and a minimum range of 6,500 m. The rockets are spin stabilised; the spin imparted by grooves in the launch tubes is sustained by offset fins which deploy after launch. The launch tubes are laid out in two banks each with 18 tubes in rows of five, six, six and one. Each tube is 3·9 m long.

The weapon is laid from a seat set between the bundles of launch tubes. The sights consist of a conventional periscope dial sight (the Panoramic Sight Type 39) and elevation scale. Laying is mechanical.

Firing control allows single rounds, part or whole salvos to be fired. The full salvo is despatched in 18 seconds; reloading takes 15 minutes with mechanical handling. Two ammunition handlers assist the detachment of three men.

At present only the anti-personnel warhead is in service. The anti-vehicle warhead will contain eight Pandora AT I mines. Under development still is the anti-armour warhead which will contain between three and five Medusa AT 2 hollow charge mines. The rocket has a high payload for its size which gives plenty of scope for the design of highly effective warheads.

The launch vehicle is the Maginus Jupiter 7·5 ton 6 × 6 truck. The cab is armour plated and contains the electrical firing control gear: this arrangement precludes firing in an arc 25° either side of the rear of the vehicle.

Two steadying jacks are manually lowered behind the rear wheels of the vehicle. All-up weight is in excess of 15,000 kg. A two wheel trailer launcher has been developed, but has yet to enter service. It weighs 19,000 kg and carries 15 tubes. The system was introduced in 1970, and each German division is scaled to two batteries of eight vehicles.

The meteorological correction of the moment for a Lars Battery is obtained by the Conar Fire Control System consisting of a radar and a computer. A single rocket is fired on the estimated range and bearing to the target which 'self-destructs' $\frac{3}{4}$ way down the trajectory. It is tracked by the radar and the computer extrapolates

The 110 mm Lars, showing the unusual position of the layer

The Conar radar used in conjunction with Lars to determine the correction of the moment

Rocket launcher BM 21

Union of Soviet Socialist Republics

	Light		Heavy
Calibre	122 mm		
Rocket length	1·905 m	or	3·23 m
Rocket weight	46 kg	or	77 kg
Fuel	Solid		
Stages	One		
Guidance	Nil		
Range	14,000 m or 20,000 m		
Warhead	HE		
Tubes/Rails	40		
Traverse limits	240°		
Reload time	10 mins		
Detachment	6		

The BM 21 is the smallest of a family of Russian multiple rocket launchers: these weapons have been a prominent feature of the Soviet order of battle since the introduction of the Katyusha rocket in 1941. The BM 21 was first seen in 1964. The single stage, solid fuel rocket is given an initial spin by the helical groves in the launch tube. Fins to sustain the spin are spring loaded to deploy after launch. The motor has a six jet nozzle, and a conventional nose fuse can be set to impact or delay action.

The BM 21 consists of a bank of 40 tubes arranged in four rows of ten, and mounted on a URAL 375 truck.

The BM 21 multi-launcher mounted on an URAL-375 truck.

The mounting is a simple turntable set over the rear wheels. A tubular extension to the left rear of the mounting carries the sight bracket: it turns down to bring the sight to a convenient height for a man standing on the ground to operate. The truck cab is retained and not protected. This lack of protection, and the downward turn of the sight bracket limit traverse to two arcs—140° to the left and left rear, and 90° to the right.

The Czech version showing the automatic loading system

It will be noted that there is both a light and heavy version of this rocket in service.

The rockets are normally fired in a battery salvo to take full advantage of their shock and saturation effects.

A single tube adaption of this rocket is used by the North Vietnamese. The 2,460 mm tube and its tripod mount are man packed as two loads of 22 kg and 28 kg respectively.

There is a Czechoslovak version of this equipment called the M 1972 consisting of the BM 21 launcher with a 40 rocket automatic reloading system mounted on a TATRA 813 lorry.

Employment

Afghanistan	German (GDR)	Syria
Angola	Hungary	USSR
Czechoslovakia	Iran	Vietnam
Egypt	Poland	

A single-tube launcher with projectile displayed below

128 mm rocket launcher RM 130 (M 51)

Czechoslovakia

Calibre	128 mm
Rocket length	800 mm
Rocket weight	24·2 kg
Fuel	Solid
Stages	One
Guidance	Nil
Range	8,200 m
Warhead	HE (24·2 kg)
Tubes/Rails	32 tubes
Traverse limits	30°
Reload time	2 mins
Detachment	6–8

This is undoubtedly the same weapon as the Yugoslavian M 63. The launch pack of 32 tubes mounted in four rows of eight is either trailer or vehicle mounted. (The Yugoslavian version is not vehicle mounted.) All 32 rockets are fired electrically in 12 seconds. They are spin stablised. The trailer version is very similar to an obsolete American equipment, the 145 mm M 21. The chief similarity is the arrangement of the launch tubes on a two wheel carriage with split trail legs, like a field gun. Conventional elevating and traversing gear is used; traverse being limited without moving the carriage. An eight man detachment is employed, most of whom are loading numbers.

The vehicle mounted version is usually seen on the PRAGA V3S 5 ton 6 × 6 vehicle. Rumania uses the ZIL 157 truck. Traverse is limited to 240°. The arrangement of the vehicle and launch tubes is similar to the Russian BM 21.

There is a similar Russian equipment—the 16 tube BM 13—firing a longer rocket. This is no longer in service, but is believed to be used as a training expedient in East Germany and Poland.

Employment	Austria	Czechoslovakia	Rumania
	Bulgaria	Egypt	Yugoslavia
	China		

The launcher mounted on a LICW Praga V-3S vehicle

The Yugoslav towed version in action

140 mm rocket launcher M-1965

Union of Soviet Socialist Republics

Calibre	140 mm
Rocket length	1,092 mm
Rocket weight	39·6 kg
Fuel	Solid
Stages	One
Guidance	Nil
Range	10,500 m
Warhead	HE (41 kg)
Tubes/Rails	16
Traverse limits	28°
Reload time	—
Detachment	—

The M 1965 is a special light weight weapon suitable for airborne forces. It is mounted on a carriage with split trails and consists of four rows of four launchers. The launcher tubes are similar to those of the BM-14, and it has the same performance as the earlier version.

Employment USSR

140 mm M-1965 launcher being loaded

The 140 mm rocket launcher mounted on GAZ 63 truck

Rocket launcher BM-14

Union of Soviet Socialist Republics

	14/8	*14/16*	*14/17*
Calibre (mm)	140	140	140
Rocket length (mm)	1,092	1,092	1,092
Rocket weight (kg)	39·6	39·6	39·6
Fuel	Solid	Solid	Solid
Stages	One	One	One
Guidance	Nil	Nil	Nil
Range (m)	10,600	10,600	10,600
Warhead	HE	HE	HE
Tubes/Rails	8 tubes	16 tubes	17 tubes
Traverse limits	30°	30°	210°
Reload time (mins)	2	3/4	4
Detachment	5	5–7	6

The BM 14 is fired from four mountings, but is always characterised by the short launch tubes. The options are:

14/8 (eight tube launcher). A trailer mount weighing 600 kg. It is used by the airborne forces of Poland and was first seen in 1963. Reload time is very short. Traverse is limited to 30°, but the elevation gives the same range as other mountings. The trailer is towed by the GAZ 63. The weapon is known in Poland as the WP 8.

14/16 (16 tube towed launcher). Known as the BM 14/16 or M 1965, this configuration consists of a 1,500 kg trailer with split trail legs like a gun. The tubes arranged in four rows of four are mounted in a simple U-shaped cradle. It has a detachment of five, and two extra loaders to keep down the reload time.

14/16 (16 tube SP launcher). This is an arrangement of 16 launch tubes in two rows of eight mounted on a ZIL 151 truck chassis.

14/17 (17 tube launcher). This seems to be the original arrangement of tubes and was first seen in 1959, mounted on a GAZ 63. Traverse is limited by the need to protect the front of the vehicle from the rocket efflux.

All varieties of this weapon are known as the BM 14, so that it is not possible to distinguish which user nations have which version.

Employment

Algeria	Germany (GDR)	Syria
China	Hungary	USSR
Czechoslovakia	Poland	Vietnam
Egypt	Rumania	

---

147 mm RAFALE weapon system

France

Calibre	147 mm
Rocket length	3,200 mm
Rocket weight	78 kg
Fuel	Solid
Stages	One
Guidance	Nil
Range	—
Warhead	19 kg Anti-Personnel (35 fragmentation bomblets) Anti-tank (56 shaped *charge* bomblets or 5 minelets)
Tubes	30
Traverse limits	—
Reload time	—

RAFALE is still under development in France and little is known about its performance except that accuracies are claimed of 1·1% of range for length zone and ·4% of range for lateral spread. This is achieved by firing a meteorological sighting missile before a salvo, which is tracked by radar to deduce the meteorological correction of the moment for each target. The 30 tube launcher is currently mounted on a Berliet lorry.

Employment Development in France

---

158 mm BREDA BR 51 GS

Italy

Calibre	158 mm
Rocket length	30/38 cm
Rocket weight	123 kg
Fuel	Solid
Stages	One
Guidance	Nil
Range	24,000 m
Warhead weight	—
Warhead	Preformed fragmenting (60 kg) Canister Incendiary Anti-armour
Tubes/Rails	10
Traverse limits	—
Reload time	—
Detachment	—

The BR 51 GS is in an advanced stage of development. Accuracy is quoted as 0·75 per cent of range. At the maximum range this produces a range spread of 180 m, thus a salvo of rockets should produce a good saturation effect. The anti-armour round is of the multiple warhead type, and increases the overall rocket length from 300 mm to 380 mm. Range is adjusted by air brakes, which are pre-set before launch and if fully deployed reduce the maximum velocity from 750 m/sec to 690 m/sec. The minimum range is unusually high—7,500 m.

A multiple launcher mounted on a tracked chassis or wheeled is planned, and each launcher will be able to fire 300 rockets per hour, from ten launch frames arranged in two rows of five.

Employment Trials only, Italy

194 mm BORA artillery rocket

Italy

Calibre	194 mm
Rocket length	4.7 m
Rocket weight	226 kg
Fuel	Liquid
Stages	One
Guidance	Nil
Range	10,000 m +
Warhead	HE (20 kg)
Tubes/Rails	3 rails
Traverse limits	—
Reload time	—
Detachment	—

The Bora is the largest of three Italian rockets which use high pressure steam to provide thrust. Like the other two members of the family (Attila II and Mira) the hypergolic fuels are pre-packed and stable when separate. Maximum speed is 390 m/sec and the range is modest for the size of the rocket. However the propulsion system has the tactical advantage of being invisible to infra-red viewers and the efflux disperses quickly. The stabilising fins have a span of 474 mm.

The warhead is described as being 'fragmentation semi-armour piercing'. An impact fuze is fitted.

The rocket is launched from a triple rail, two wheel trailer. The trailer is a triangular frame with a towing eye at the apex. The launch rails are built on a framework of tubular members. Traverse is limited, but the whole trailer can be easily manhandled through large switches in azimuth. Elevation is by twin jack pillars beneath the launch rail framework.

Note. The Italians are developing another member of this family called FLIT: calibre 152 mm, same propulsion system, 15 kg semi armour piercing HE fragmentation warhead, range, 10,000 m.

Employment Trials only, Italy

194 mm Bora on single rail launcher about to be test fired

Battlefield missile 'Wolf'

Israel

Calibre	150–200 mm?	
Rocket length	—	—
Rocket weight	—	—
Guidance	Nil	Nil
Range	1,000 m	4,500 m
Warhead	HE (170 kg)	(HE (70 kg)
Tubes/Rails	—	—
Traverse limits	—	—
Reload time	—	—
Detachment	—	—

There are two new Israeli battlefield rockets which are both called Wolf (Ze'ev). They were used in the 1973 Yom Kippur War. They are not accurate but reputedly effective against large concentrations of men or materiel. It would be logical to use a multiple launcher for a weapon with such characteristics. It is believed that a cage-type launcher is used.

There is no indication of the size of the rockets, but the weight of the warheads suggests a calibre of 150/200 mm.

Employment Israel

200 mm rocket launcher BMD 20

Union of Soviet Socialist Republics

Calibre	200 mm
Rocket length	3·11 m
Rocket weight	194 kg
Fuel	Solid
Stages	One
Guidance	Nil
Range	18,000 m
Warhead	HE
Tubes/Rails	4 frames
Traverse limits	20°
Reload time	6–10 mins
Detachment	6

This is a very elderly equipment which has been relegated to a training role by Warsaw Pact countries. It is, however, believed to be active in North Korea. It is is mounted on either the ZIL 151 for ZIL 157 trucks.

Employment North Korea
Warsaw Pact countries (training only)

200 mm rocket launcher BMD 20

203 mm general support rocket system

United States of America

This project began in September 1977 and involves a tracked vehicle (based on the XM 723 Fighting Vehicle chassis) carrying a rotatable launcher, similar in general concept to the German Lars. On to this launcher two pre-loaded ready-to-fire pallets containing 6 rockets each will be placed, so that reloading becomes simply a matter of removing an empty pallet and replacing it with a loaded one, an operation estimated to take 5–8 minutes when performed from a special support vehicle.

The rockets, of 203 mm diameter, are to be about 4 m long. The standard warhead will be a combined hollow-charge and fragmentation unit to give useful effect against hard or soft targets. Future designs envisaged include anti-tank, smoke and chemical warheads, while there are also suggestions of carrier warheads to deliver scatterable mines and anti-armour bomblets.

At the time of writing the first design prototypes are undergoing evaluation.

216 mm E-3 rocket

Spain

Calibre	216 mm
Rocket length	1,406 mm
Rocket weight	101 kg
Fuel	Solid
Stages	One
Guidance	Nil
Range	14,500 km
Warhead	HE (37·5 kg) Incendiary Smoke
Tubes/Rails	21 cages
Traverse limits	90°
Reload time	14 mins
Detachment	—

The E-3 rocket belongs to the same family as the R-6-B-2. It too has an IMI solid fuel motor, is spin stabilised and has a quoted delivery error of one per cent of range. The rockets are fired electrically at two second intervals.

The 21 launch cages are mounted on a pedestal which is laid manually. Elevation of the launching assembly is by screw jack.

The pedestal and launch cages are mounted on a Banieros Pante III truck—the L-21/E-3 launcher. The vehicle is a 6 × 6 wheeled chassis with an open cab and platform body. A steady jack is lowered to provide a stable firing platform: it consists of a framework of tubular members which hinges up when not in use.

A trailer mounting is believed to exist.

Employment Spain

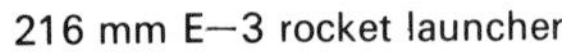
216 mm E—3 rocket launcher

240 mm rocket launcher BM 24

Union of Soviet Socialist Republics

Calibre	240 mm
Rocket length	1·180 m
Rocket weight	113 kg
Fuel	Solid
Stages	One
Guidance	Nil
Range	11,000 m
Warhead	HE
Tubes/Rails	12 cages
Traverse limits	210°
Reload time	3–4 mins
Detachment	6

This stubby rocket is launched from an arrangement of two rows of six cages mounted on the ZIL 151 truck. The layout is very similar to that of the BM 21, including the sight mounting. In this case, however, the cab is protected by folding steel screens which cover the windows. Steadying jacks are fitted behind the rear wheels.

The Soviet Army has a version mounted on the AT-S medium tracked chassis, but this has yet to be exported.

Employment Algeria, Egypt, Germany (GDR), Israel, Poland, Syria, USSR

The BM-24 deployed. Note blast shields on windscreen of truck

250 mm rocket launcher BMD 25

Union of Soviet Socialist Republics

Calibre	250 mm
Rocket length	5·822 m
Rocket weight	455 kg
Fuel	Solid
Stages	One
Guidance	Nil
Range	30,000 m
Warhead	HE
Tubes/Rails	4 cages
Traverse limits	—
Reload time	—
Detachment	—

BMD 25 on JAAS-214 truck

The BMD 25, first seen in 1957, is a large rocket, fired from a cage. The four fins are permanently deployed and engage in helical rails in the launch cage to impart spin. The four cages are arranged in a single row, mounted on a ZIL 151 truck. KrAZ 214 is an alternative chassis. The cab is projected by folding screens and rear steadying jacks are fitted. The arrangement is exactly that of the BM 24.

A version with six launch cages is believed to exist.

Employment USSR

300 mm 30 rocket

Japan

Calibre	300 mm
Rocket length	4·5 m
Rocket weight	—
Fuel	Solid
Stages	One
Guidance	Nil
Range	25,000 m
Warhead	—
Tubes/Rails	2 rails
Traverse limits	—
Reload time	—
Detachment	—

The Type 30 rocket is the largest of a family of rockets produced by the Nissan Motor Company's Space and Aeronautical Department. No details of other members of this family are available. This weapon has an excellent range for its size. Small cruciform fins are fitted and the rocket appears to have spin motors.

The general arrangement of the Hino launch vehicle is similar to the Honest John, except that there are two launch rails. There are three hydraulic steadying jacks, one at the rear, and one at each side forward of the 8-wheel twin rear axles. Like the Honest John launcher there is no driver's cab and traverse is limited.

Employment Japan

---

300 mm D-3 rocket

Spain

Calibre	300 mm
Rocket length	1·898 m
Rocket weight	247·5 kg
Fuel	—
Stages	One
Guidance	Nil
Range	17,700 m
Warhead	HE (89·4 kg) Smoke Incendiary
Tubes/Rails	10 cages
Traverse limits	90°
Reload time	—
Detachment	—

The D-3 rocket is a larger version of the E-3. It is claimed to have the same delivery accuracy—one per cent of range. The rate of fire is the same for both equipments—one round in two seconds.

The rockets are launched from cages, which are only 22 mm longer than the rocket. Ten cages arranged in two rows of five are mounted on the Banieros Pante III truck to make the L-10/D-3 launcher. The pedestal mount and screw jack elevating gear are from the E-3 system.

A mechanical re-loading system is used.

Employment Spain

300 mm D-3 being fired from L-10/D-3 launcher

381 mm G-3 rocket

Spain

Calibre	381 mm
Rocket length	2·657 m
Rocket weight	527·5 kg
Fuel	Solid
Stages	One
Guidance	Nil
Range	23,500 m
Warhead	HE (217 kg)
Tubes/Rails	8
Traverse limits	45°
Reload time	—
Detachment	—

Little is known about the G-3 rocket that is not tabulated above. It is the largest of the Spanish rockets, and boasts a very large warhead. It is launched from a short cage, which is only 163 mm longer than the rocket. Accuracy of one per cent of range is claimed.

The L-8/G-3 launcher is self propelled and wheeled.

Employment Spain

557 mm battlefield missile Lance

United States of America

Calibre	557 mm
Rocket length	6·17 m
Rocket weight	1,530 kg
Stages	One
Fuel	Liquid
Guidance	Inertial
Maximum speed	—
Range	120 km
Warhead	Nuclear or conventional
Tubes/Rails	One
Traverse limits	—
Reload time	—
Detachment	6

American battlefield guided missiles trace their ancestry back to the German V2 rockets of World War II. The first was Corporal, then Sergeant and now Lance, which also replaces Honest John and thus simplifies the order of battle. Basic research started in 1959: first firings were in 1965 and the weapon entered service in 1972.

The missile is made up of three basic assemblies: warhead, main rocket assemblage and control surfaces. The main rocket assemblage contains the guidance module, propellant tanks and rocket motors. The propellant consists of two hypergolic agents: an oxidiser (inhibited red fuming nitric acid), and the fuel (unsymmetrical dimethylhydrazine).

Advanced storage techniques combine the convenience of storage and handling associated with solid propellants with the power and fine thrust control possible with liquid fuels.

At launch a boost motor accelerates the missile to the required speed over a period of between 1·5 and 6 seconds: it then shuts down and a sustainer motor continues to propel the rocket in a state of zero acceleration. The duration of the boost depends on the required range. An accelerometer determines the point at which the boost motor is shut down, and then controls the infinitely variable thrust of the sustainer motor.

Lance and launcher on wheeled mountings

The weapon is stabilised initially by spin which is imparted by tapping off gases from the solid propellant gas generator (the 'starter motor' for the main motors). This system is active for 1·5 seconds, and spin is maintained by the fins. Thrust-vector control steers the missile during flight.

The inertial guidance system is immune to all known enemy counter measures. It automatically compensates for meteorological influences and gives the missile a 'fire and forget' characteristic. The missile is laid in azimuth like a gun. Target range data are fed to the missile before firing by a monitor programmer. This solid state device also supplies pre-launch electrical power, monitors the firing sequence and controls the arming of the firing mechanism, yet it is only 42 × 51 × 43 cm.

Lance can be fitted with the M 234 nuclear warhead or can carry conventional payloads such as the M 251 cluster-bomb type.

The compact rocket is launched from a tracked vehicle derived from the M113 AFV. The engine and running gear are unchanged except for the addition of suspension locks to provide a stable launch platform. Traverse, and elevation to the firing angle, are manual. The vehicle with its six man detachment is capable of 64 km/hr and swims at 10 km/hr.

The launch rail assembly can also be mounted on a light two wheel trailer weighing 3,000 kg complete with rocket. This can be air dropped, and gives air delivered and air mobile forces a nuclear weapon.

The launcher vehicle is accompanied by a loader/ transporter carrying two rockets. It is fitted with a hydraulic handling unit or crane which both loads the launcher and handles the mating of warhead to motor. Control surfaces in the shape of four rear mounted fins are fitted at the launch site during deployment.

Employment

Belgium	Netherlands
Germany (FRG)	UK
Israel	USA
Italy	

557 mm Lance missile and launcher mounted on M 113

650 mm Pluton

France

Calibre	650 mm
Rocket length	7·64 m
Rocket weight	2,350 kg
Fuel	Solid
Stages	One
Guidance	Inertial
Range	120 km
Warhead	Nuclear (500 kg)
Tubes/Rails	One
Traverse limits	N/A
Reload time	—
Detachment	4

Issue of Pluton to five regiments, each of six launchers, began in 1975. The missile is delivered as three major components: motor, warhead and guidance module. The motor is single stage, solid fuel. The weapon is assembled onto the launch vehicle using the motor's delivery container as the launch tube.

The rocket flies free for the first five seconds before the pre-programmed guidance system takes over. Changes in course are effected by control surfaces at the tips of the fins. After launch the onboard computer adjusts the flight path using inputs from an accelerometer and gyro reference. A simple, but effective back-up inertial guidance system is mounted in parallel to the primary system. This second system serves to monitor the weapon's performance and acts as a safety device. Before the warhead is initiated a comparison of data between both guidance systems must agree within predetermined limits.

The launch position command post is equipped with an IRIS 35M computer. This produces target data for direct input to the missile guidance system based on firing position survey data and target coordinates. A second IRIS 35M computer mounted on the launcher vehicle monitors the weapon and firing sequence and automatically carries out pre-flight checks. It also accepts orientation for the missile from a north seeking gyro.

The warhead is the AN 51 which is said to have 10 kiloton and 25 kiloton options.

The weapon systems reaction time is 30 minutes.

The missile launch vehicle is based on the AMX chassis, which runs on five road wheels with five return rollers. It is powered by the Hispano Suiza HS 110 Flat-12 cylinder diesel, which is multi-fuelled, supercharged, and produces 720 bhp. The centrifugal clutch is operated electrically by movement of the mechanical five speed gear change. The main engine is supplemented by a small turbine which provides pre-launch power for the missile and launcher functions. Protection of the crew from nuclear and chemical attack is provided using the same air filtration and pressurisation equipment as the AMX 30 tank.

A dual purpose hydraulic crane is fitted over the right hand track. It handles the weapon assembly and when laid flat extends under the launch container as the means of elevation. There is no provision for laying in azimuth; placing the vehicle within 10° of the line of fire is sufficient for the guidance system to be effective.

Employment France

Pluton missile in its container/launcher ready to deploy. For firing, the container/launcher elevates to the vertical

762 mm Honest John

United States of America

Calibre	762 mm
Rocket length	7·5 m
Rocket weight	2,040 kg
Fuel	Solid
Stages	One
Guidance	Nil
Range	37,000 m
Warhead	Nuclear HE
Tubes/Rails	One
Traverse limits	4°
Reload time	—
Detachment	4

Honest John is a simple, unguided surface-to-surface rocket, designed to be fitted with nuclear or conventional warheads. It has a single stage, solid fuel motor. Supersonic speed is reached during the four second motor burn time.

The missile is stabilised in flight by four tail fins. Two spin motors ignite as it leaves the launcher and neutralise any eccentricity in the motor performance. Accuracy throughout its range band (7·5 to 37 km) compares favourably with an equivalent heavy artillery piece.

Sighting is by stardard clinometer and dial sight, mounted at the left rear of the launch vehicle. Laying is manual, but top traverse is limited to 4°: thus great care must be exercised in deploying the launcher for its target.

Developmental policy has been one of continual improvement. No modifications have greatly changed the shape or size: it is still 7·5 m long and weighs over 2,000 kg at launch. After 15 years operational service,

762 mm Honest John rocket elevated ready for firing, but note jacks on vehicle have not yet been lowered

production has stopped, and most but not all users are phasing Honest John out of their order of battle in favour of Lance.

The warhead, motor, and fins are assembled together into a complete missile in the field. Once assembled, the missile is loaded onto the 6 × 6 International Harvester launch vehicle. This vehicle has no cab superstructure: the windscreen folds flat during firing to avoid damage. Four steady jacks are deployed to provide a stable launch platform. The vehicle has excellent battlefield mobility, both loaded and unladen.

The launching party consists of the launcher, a command post vehicle and a windset trailer which measures the low level wind at launch.

Simplicity is the great virtue of the whole weapon system. It leads to robust and reliable equipment, and requires few techniques which are not common with conventional guns. It is now being replaced by Lance.

Employment	Denmark	South Korea	Turkey
	Greece	Taiwan	

FROG missile system

Union of Soviet Socialist Republics

'FROG' (Free Rocket Over Ground) is the NATO designation for a family of unguided Russian missiles. The Russians name may be LUNA. Each of the six types represents a continuing development programme, and the rocket diameter is not constant: for convenience all FROGS have been grouped together.

FROG is similar to Honest John—the nearest equivalent Western weapon. They are free flight missiles built round a solid fuel motor. Stabilisation is by spin which is sustained in flight by fins. Like Honest John, the missiles are fired from a launcher/transporter.

The FROG family are all nuclear weapons. However, the size and payload could allow conventional multiple warheads to be fitted. The HE warheads used by Egypt and Syria in the Yom Kippur War were ineffective because the delivery error exceeded the radius of damage.

In the Soviet Army FROG is deployed with tank and motor rifle divisions in battalions of four launchers.

Employment

Algeria (4)	Germany (GDR) (7)	Poland (7)
Bulgaria (7)	Hungary (7)	Syria (7)
Cuba (4)	Iraq (7)	USSR (1–7)
Czechoslovakia (7)	North Korea (5 & 7)	
Egypt (3 & 7)	Rumania (7)	

FROG 1

Union of Soviet Socialist Republics

Calibre	850 mm
Rocket length	10·0 m
Rocket weight	3,000 kg
Fuel	Solid
Stages	One
Guidance	Nil
Range	25–65 km
Warhead	Nuclear HE Chemical
Tubes/Rails	One
Traverse limits	—
Reload time	—
Detachment	—

FROG 1 is an obsolete weapon, first introduced in 1957. It is launched from a ribbed box container. This could, like the French PLUTON system, be the delivery container for the motor section of the missile. The container certainly forms a heater jacket to maintain a constant temperature for the solid fuel. The six-fin tail, and the warhead project out of the container at either end. A tubular framework protects the warhead which overhangs the vehicle when travelling and is of much greater diameter than the motor body.

The missile is transported on, and fired from, a vehicle based on the JS 3 tank chassis. It has six road wheels, and three return rollers. The engine is a V 12 diesel which produces 520 bhp. All-up weight is in the order of 36,000 kg.

The built-up superstructure of the vehicle incorporates an enclosed cabin. The carrying cradle is also the launch rail, and is elevated by a hydraulic ram. There are steadying jacks at the rear so that the suspension does not lock for firing as it does with newer Western equipments. Traverse is probably very limited.

FROG 2

Union of Soviet Socialist Republics

Calibre	600 mm
Rocket length	9·0 m
Rocket weight	2,400 kg
Fuel	Solid
Stages	One
Guidance	Nil
Range	25 km
Warhead	Nuclear
	HE
	Chemical
Tubes/Rails	One
Traverse limits	—
Reload time	—

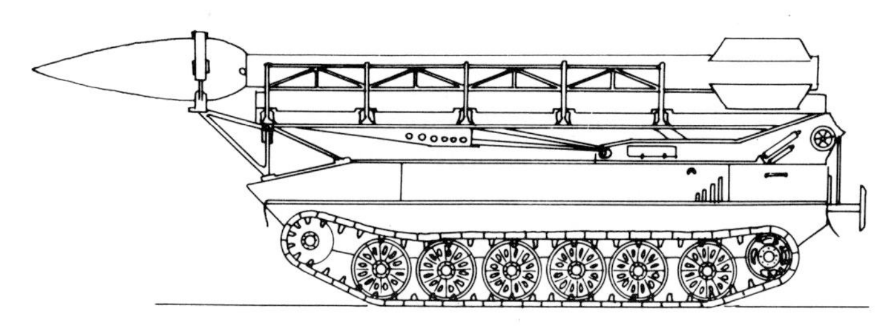

FROG 2

FROG 2 is a lighter, smaller rocket than its predecessor. It has a simpler cruciform tail fin assembly and the warhead is still of greater diameter than the motor section. The missile is carried on a framework of U-shaped ribs and longitudinal side members. A tubular frame supports the front of the launch rail for travelling. An 'A' Frame under the launch rail is used to elevate the rail and missile to the launch angle. Split function laying is used.

The vehicle is the tracked PT 76 chassis with six road wheels and no return rollers. It has a six cylinder diesel engine developing 240 bhp. The superstructure of FROG 1 is missing and the careful designing has resulted in a weight of only 15,000 kg. The PT 76 is an inherent swimmer, riding high in the water propelled by hydro-jets. The FROG 2 mounting may well possess this quality also. It has a road speed of 35 km/hr, and an endurance of 250 km.

FROG 3

Union of Soviet Socialist Republics

Calibre	550 mm
Rocket length	10·5 m
Rocket weight	2,280 kg
Fuel	Solid
Stages	Two
Guidance	Nil
Range	40 km
Warhead	Nuclear (250 kg)
	HE
	Chemical
Tubes/Rails	One
Traverse limits	—
Reload time	—
Detachment	—

When introduced in 1960 the FROG 3 was the first rocket of its type to have two stages. The first stage is fitted with four stabilising fins, in line with those of the second stage. The first stage motor is made up of 12 jets built around a central nozzle. The warhead is less shaped than its predecessor, consisting of a simple cylinder and conical nose cone.

The vehicle and mounting are exactly the same as FROG 2, except that two track return rollers are fitted above the road wheels.

FROG 3 being prepared for launching

FROG 4 and 5

Union of Soviet Socialist Republics

Calibre	400 mm
Rocket length	10·2 m
Rocket weight	2,040 kg
Fuel	Solid
Stages	Two
Guidance	Nil
Maximum speed	—
Range	45 km
Warhead	Nuclear (300 kg) HE Chemical
Tubes/Rails	One
Traverse limits	—
Reload time	—
Detachment	—

Although the range is quoted as much greater than FROG 3, these weapons appear to be identical to it except in the shape of the warhead, which is a simple cone fitting flush with the motor body. FROG 5 has a further modification to the nose cone, but is otherwise a FROG 4. The figures quoted for warhead weight are greater than for FROG 3.

It is believed that FROG 4 and 5 launchers may have a built-in north-seeking gyro to assist in laying.

FROG 5 being prepared for firing: note conical nose section

FROG 4

FROG 6

Union of Soviet Socialist Republics

FROG 6 is a training vehicle with dummy rocket. It has rarely been seen in public. Its production may have been made necessary because of the increasing complexity of the sighting systems attributed to FROGs 4 and 5.

FROG 7

Union of Soviet Socialist Republics

Calibre	600 mm
Rocket length	9·0 m
Rocket weight	6,300 kg
Fuel	Solid
Stages	One
Guidance	Nil
Maximum speed	—
Range	60 km
Warhead	Nuclear HE Chemical
Tubes/Rails	One
Traverse limits	—
Reload time	—
Detachment	—

FROG 7 has probably been in service since 1965, but was first shown to the public in 1967. It is the latest of the FROG range of missiles and the first to have a wheeled launch vehicle.

It is altogether bigger than preceding types. It reverts to being a single stage rocket. The main motor venturi is surrounded by a ring of 20 smaller nozzles. The warhead is flush with the motor body, and the fins are both narrower and longer.

The launcher/transporter is the eight wheeled ZIL 135. Steadying jacks fold down to provide a stable firing platform. For the first time, the missile sits on its launch rail (which is of triangular cross section like Honest John's), and not in a cradle. For travelling, the rocket is restrained by two straps. The cab of the ZIL 135 is a conventional lorry type, suggesting that some sort of screen is used to protect the glass when the missile is launched.

The nuclear warhead is in the 25 kiloton class.

A tender based on the ZIL 135 carries three missiles ready and prepared for firing. A crane vehicle is required to transfer them to the launch rail.

Above: FROG 7 unit exercising. Note that two launchers are deployed together

Left: ZIL 135 truck carrying three FROG 7 rockets

FROG 7 showing jacks in position and laying mechanism

Battlefield missile—SCUD A (SS-1B)

Union of Soviet Socialist Republics

Calibre	850 mm
Rocket length	10·5 m
Rocket weight	4,500 kg
Fuel	Liquid
Stages	One
Guidance	Radio command
Range	150 km
Warhead	Nuclear
Tubes/Rails	One
Traverse limits	N/A
Reload time	—
Detachment	—

SCUD is the NATO name for this huge rocket which was first seen in 1957. It is launched vertically, and guided by radio command during flight. Course corrections are effected by control surfaces at the rear of the motor, which also act as stabilising fins. The launch vehicle is basically that of FROG 1. Based on the JS 3 tank chassis, it has a built-up superstructure which contains a cabin for the detachment.

The missile is carried in a frame of tubular members which extends forward round the nose cone to protect it. This frame raises the rocket to its vertical launch position behind the chassis, and provides a ladder for access to the warhead during final check-out procedures. It is lowered for firing.

Although details of the launch procedure have not been released, the fuelling and guidance check-out must involve a considerable number of men and vehicles at the launch site. The weapon is most unlikely to travel after fuelling. The radar and radio vehicles needed by the type of guidance used increases the size of the firing unit. Reaction time is claimed as one hour, and the warhead is reputedly in the 100 kiloton class.

The following list of users refers to SCUD A and B as no definite indication is available.

Employment			
	Bulgaria	Hungary	Syria
	Czechoslovakia	Iraq	USSR
	Germany (GDR)	Poland	
	Egypt	Rumania	

(illustrated overleaf)

SCUD A on wheeled launchers ready for firing

SCUD A on tracked launchers coming into action

Battlefield missile SCUD B (SS-1C)

Union of Soviet Socialist Republics

Calibre	850 mm
Rocket length	11·0 m
Rocket weight	6,300 kg
Fuel	Liquid
Stages	One
Guidance	Inertial
Range	165–280 km
Warhead	Nuclear
Tubes/Rails	One
Traverse limits	N/A
Reload time	—
Detachment	—

SCUD B

SCUD B is a direct development of SCUD A, and considerably heavier. It continues to use the tail fins as control surfaces. Its fuel is described as 'storable'.

It was first seen on the SCUD A tracked launch vehicle. The new vehicle is an eight wheel MAZ 543, first seen in 1965. The nose of the missile, protected by a single tubular rail, sits between two cabs: one for the driver, the other for the missile control console and operator. The body of the vehicle contains the detachment and control gear. The launch pad is integral to the erector frame. The rear of the vehicle is supported on jacks while the missile is elevated to the vertical for launching.

The missile is held in the erector frame by clamps behind the warhead.

SCUD is deployed in brigades of nine launch sections and attached to fronts.

There have been rumours of a SCUD C weapon, but it is believed that these arise from confusion over the two launch vehicles used for SCUD B.

Employment Germany (GDR)
USSR

Battlefield missile—Scaleboard (SS-12)

Union of Soviet Socialist Republics

Calibre	850 mm +
Rocket length	11·0 m +
Rocket weight	—
Fuel	Liquid
Stages	One
Guidance	Inertial
Range	700–800 km
Warhead	Nuclear
Tubes/Rails	One
Traverse limits	N/A
Reload time	—
Detachment	—

Scaleboard is the largest Soviet battlefield missile, and was first introduced to the public in November 1967.

The missile is transported in a heavily ribbed container which is only removed after erection to the vertical launch position behind the erector/launcher vehicle. It is believed that the motor is liquid fuelled, and that the warhead is in the megaton class, but few facts about this important weapon have been released.

The container and missile are transported on a MAZ 543 wheeled erector/launcher similar to that used for SCUD A and FROG 7.

Employment USSR

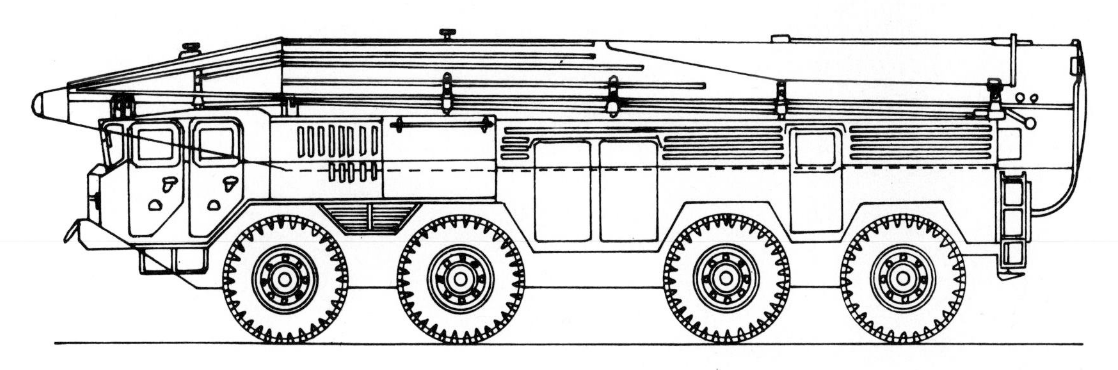

Scaleboard (SS-12)

Battlefield support missile Pershing (MGM 31)

United States of America

Calibre	1,000 mm
Rocket length	10·6 m
Rocket weight	4,600 kg
Fuel	Solid
Stages	Two
Guidance	Inertial
Range	160–750 km
Warhead	Nuclear
Tubes/Rails	One
Traverse limits	N/A
Reload time	—
Detachment	—

The original Pershing was brought into service in 1962, and was deployed to Europe in 1964. With the introduction of an improved version in 1969, the two types became known as 1 and 1A. Pershing 1 is no longer in service.

The first stage motor is fitted with four tiny triangular stabilising fins. The second, sustainer stage has four control surfaces.

The Pershing 1A is carried on an articulated M 656 wheeled erector/launcher, which is fitted with a launch pad for the vertical take-off. The rocket can be carried complete after warhead mating—an improvement over the earlier model. The launch equipment consists of:

(a) Erector/launcher.
(b) Programme test and power generator on a second M 656.
(c) Control truck.
(d) Radio terminal vehicle with pneumatically raised aerial.

An automatic count-down process has reduced reaction time but the rocket is unchanged for 1 and 1A systems.

The whole system is airportable in C-130 Hercules aircraft, and is operated in battalions of four firing batteries.

Employment Germany (FRG) USA

The Pershing missile after warhead mating, ready for launch

VII. Ancillary Equipment for Surface-to-surface Artillery

As artillery weapons have become more sophisticated, so they have become dependent on a variety of ancillary equipments in order to realise their full potential. A field gun firing in the simplest indirect fire role has always required an observer, communications, a command post equipped to produce gun-sight data, and an adequate supply of ammunition. The more sophisticated and longer range systems of today require in addition the means of acquiring targets 'well over the hill'; night observation aids, range finders and other survey instruments for their observers; and the means of measuring meteorological changes and muzzle velocities. Furthermore they need computers to assess and collate the mass of target information and to provide almost instantaneous firing data at the guns and rockets tasked to engage each target. In this chapter some of the specialist equipments used by artilleries today are described in six categories: target acquisition and locating, observation post aids, survey instruments, meteorological instruments, muzzle velocity measurement, and artillery computers.

1. TARGET ACQUISITION AND LOCATING

The basic artillery problem is to achieve accurate location of the target and the guns in relation to each other. This is a reasonably simple problem, when the target can be seen from a ground observation post, but the acquisition and accurate locating of targets deep in enemy held territory, and well out of sight of ground observers is more difficult. The devices used to provide information about this kind of target fall into two categories. First, those which provide general surveillance and hence target acquisition in depth; these include drones or remotely piloted vehicles and long range surveillance radar. Secondly those which locate enemy guns, mortars and rocket launchers; this includes flash spotting, sound ranging and mortar/rocket locating radars.

Drones

AN/USD (Midge)

Canada

The AN/USD 501 is a reliable airborne reconnaissance system which flies preplanned missions out to 150 km. The drone can have either a camera or an infra-red line-scan sensor which permits detection, recognition and identification of small, low contrast ground objects. The system includes a supply of re-usable drones and facilities for flight planning and handling, launching and recovering the drones. Photographic processing and interpretation can also take place at the drone troop location. Terminal guidance and recovery is effected by a radio beacon, which causes a parachute to open for a safe descent at the troop position.

It has some advantages over manned aircraft reconnaissance in that its flight response time is short, it can operate from unprepared sites, and it is so fast and has so small a radar echoing area that it is relatively immune to anti-aircraft defences. However it is not able to send back information in flight, its mission has to be preplanned and cannot be altered once it has been launched, its sensor payload is limited, and, most important of all, its target information response time is slow, because its photographs can only be interpreted after the mission has been completed. Its detailed characteristics are still classified.

Employment Belgium
Canada
UK

The Midge being launched

R 20

France

R20 is a monoplane drone, propelled by a Marbore II jet engine. It is launched by means of a solid fuel propelled cradle from a modified standard 3–4 ton vehicle. It is guided from the ground by a tracking and control radar when it is within range, but for the major part of its course it is controlled by its own computer and gyroscopic controls. Its sensors include camera for day and night photography, infra-red line-scan, and radio activity detectors (cyclope).

Characteristics

Length	5·7 m
Wing space	3·7 m
Weight	850 kg
Engine thrust	400 kg
Speed Mach	0·65
Radius of action	160 km
Overflight accuracy	300 m
Guidance	Preset inertial

Employment France

The R-20 French drone

Developments

It is likely that remotely piloted vehicles, which can be controlled during flight and which can send back real-time target information by television, will eventually replace the drone in its target acquisition role.

Surveillance Radars

The problem with a radar system is that it can only detect moving targets in the line of sight. Its application for target acquisition in depth in broken country, such as is normal in North West Europe, is therefore limited. It does, however, have the advantage of being unaffected by weather and extreme climatic conditions. Two examples are:

RATAC

France

This reliable light radar is now in full production for the French Army. It has been designed primarily as a target acquisition radar, but it can also be used for the direction of artillery fire, guiding reconnaissance patrols and aiding helicopter navigation. Its details are still classified.

AN/TPS-58

United States of America

This is a vehicle mounted system weighing 3,500 pounds. It has an accuracy of 50 m with a detection range of 12 km for personnel and 20 km for vehicles.

Locating

There are only two methods of locating enemy guns and rocket launchers in service at the moment—flash spotting and sound ranging. Flash spotting provides the location of the gun by plotting the bearings to its muzzle flash as seen from a number of observation posts. Sound ranging uses a device to measure the difference in time of arrival of the sound of discharge at a number of microphones whose locations are fixed by survey.

It has always been possible to find the positions of enemy guns by their muzzle flash and this still remains the most simple method of locating. Observations are made from a number of accurately surveyed posts using a theodolite-type instrument to the gun flash. The coordinates of the gun can then be determined by computation or graphical intersection. There are however practical difficulties in ensuring that all observers record the same flash and that bearings are taken with sufficient accuracy to produce dependable results as the batteries are usually sited in defilade and direct observation is impossible. These can be overcome by a high standard of observer training. Flash spotting has been a generally neglected art in many armies in recent years, but the Egyptians demonstrated in 1973 that it is still a worthwhile system.

Sound ranging was first used effectively in World War I and has the advantage over flash spotting in that it can locate well beyond the field of view of an observer and it is not affected by the visibility. A series of accurately surveyed microphones is established across the front at intervals of 300–700 m for 'short base' locating, and 1,000–2,000 m for 'long base' locating. The whole system is called a 'sound ranging base', and a pair of microphones is termed a 'sub-base'. The time of arrival of the sound of an enemy gun or rocket launcher firing is recorded electrically by each microphone and, since each pair of microphones constitutes two known points with a known time interval, the direction of the explosion from each sub-base can be calculated. In practice the recordings of at least three microphones are needed to give an accurate plot, and more are desirable at longer ranges. It is a fairly simple matter to design a plotting board which, having allowed for meteorological corrections, will show the location of an enemy gun at the intersection of the bearings from each microphone. Accurate location of guns can be obtained by this method out to 1½ times the base length for a Long Base (e.g. out to 15,000 m for a 10,000 m base) and out to 2½ times the base length for the Short Base (e.g. out to 3,750 m for a 1,500 m base).

Mortar Locating Radars

These radars operate by 'observing' the flight of a hostile projectile and extrapolating its trajectory back to the source. The mortar bomb's fin is an excellent radar reflector and its high trajectory brings it above the horizon early in its flight before any ballistic variables have had time to influence it. The vertical attitude of the bomb in flight presents the optimum reflecting shape, and the steep trajectory enables several observations needed to determine its shape to be clearly separated in elevation. Only mortars can be located at present, but the extension of this concept to the location of multiple rocket launchers is a future possibility. A satisfactory radar able to track the far flatter trajectories and smaller echoing areas of shells from guns is not yet in service although one is believed to be under development in the United States.

Cymbeline

United Kingdom

Cymbeline is a rugged lightweight portable locating radar intended primarily for mortar location and for the adjustment of artillery fire. It is a completely self-contained radar system, including is own power supply, with a detachable display unit. The radar is mounted on a four-legged structure supported on screw jacks fitted with hydraulic shock absorbers.

The antenna system consists of a Foster scanner which illuminates a parabolic cylinder reflector and produces a pencil beam scanning in azimuth. The complete radar head can be rapidly rotated to cover any required sector; for example, 180° rotation takes only 15 seconds. When in transit the reflector folds down.

Below the antenna is an equipment box which houses the main radar unit, the power unit and the display unit during transit. The main radar unit contains the transmitter/receiver and the radar timing and computer modules.

The display unit can be removed from the equipment box for remote operation for distances up to 15 m. It consists of a short-persistence 'B' scope on which the radar returns are displayed. It also carries all the controls necessary for the operation of the radar. The mortar co-ordinates are shown on another unit which is detached and can be used at distances up to 2 m from the display.

The radar can be operated from any 1·5 kVA dc supply at 26 V ± 4 V. However, provision has been made for a Wankel gasoline engine-driven generator to be mounted on the equipment.

The radar enables the operator to plot two points in

the bomb trajectory, and to measure the slant range and bearing to each of these positions. The time taken for the bomb to travel between the two points is also measured. The computer uses this information together with the pre-set elevation angles to determine the firing position of the mortar. This entire process takes place in about half a minute.

Additional facilities have been provided to ensure the maximum accuracy of location and ease of operation over a wide range of operational conditions. For maximum range performance a switched single beam is used, and an additional beam position is available to alert the operator for making the first intercept. For shorter range working a double-beam mode of operation may be selected to obviate operator reaction time errors. This also improves multiple target capability.

Provision has also been made for the fitting of an optional microelectronic digital data storage module. This enables the radar returns to be stored to provide a long-persistence display so that operator concentration can be reduced while improving the marking accuracy. Data storage also improves multiple handling capability.

Two operators are normally required, but in an emergency one will suffice.

Employment UK

A trailer-mounted Cymbeline mortar locating radar. In the background is the FV 432 version shown with the radar in the transit position

AN/MPQ-10A

United States of America

This is now an elderly and cumbersome equipment which weighs 2,700 kg. It uses a parabolic reflector. The AN/MPQ-10A operates from 460 m to a maximum range of 18,300 m. It scans up to 45° in azimuth.

The radar operates in conjunction with a mortar-locating computer and produces the co-ordinates of hostile mortar positions or the point of impact for bombs being tracked. There is a visual display of information.

AN/MPQ-4

United States of America

AN/MPQ-4 is similar to the British Green Archer System, and consists of two two-wheel trailers. One carries the radar head and associated circuitry, while the other carries a power generator. The primary trailer weighs 2,300 kg.

The maximum detection range is 10 km, at which range an accuracy of 50 m is claimed. It scans 25° in azimuth.

The radar is operated by one man at the radar trailer, or with the control console remoted.

It is likely to be replaced by the AN/TPQ-36 for which data is not yet available.

SNAR-2, Small Yawn, Porktrough

Union of Soviet Socialist Republics

Little is known about these three radars, which are used for mortar and to a certain extent rocket location. They are all mounted on an AT-L tracked vehicle and use the X-band. The effective range of SNAR-2 is thought to be about 9,000 m.

Employment Warsaw Pact countries
Egypt
Syria

2. OBSERVATION POST AIDS

The basic aids for an artillery observer are of course his binoculars, his map and his compass. Of the other aids towards achieving accuracy of target location and the swift application of artillery fire on to it three must be given special attention, as they have brought significant advances both in observer efficiency by day and night, and also in closing the survey loop between the guns and the target. These are the laser range finder, the night observation device and the short range radar.

Artillery range finder DS-2

Union of Soviet Socialist Republics

The Soviet range finder is displayed as part of a field artillery observation post: note map boards and aiming circle in the background.
The DS-2 is a conventional split image optical range finder. The arms can be angled upwards so that the operator can remain under cover.

Laser Range Finders

Accurate range and azimuth measurement by the observer is an essential ingredient of accurate target location. The artillery laser range finders are in general small, compact and of rugged construction. By adding the ability to measure angles the one equipment can be used to establish the co-ordinates of its own and target positions within 5 m. Where data links are provided to fire control computers, the computer can resolve the bearing and distance calculation to establish target co-ordinates. Some examples of laser range finders in service are given below.

VAL 11501 artillery laser

Multi-National (Sweden and Yugoslavia)

Sweden and Yugoslavia have co-operated to produce the VAL 11501. It is related to the VAL 11210 tank laser. Both use a neodynium glass rod laser medium, and the main difference lies in the range: the artillery laser measures distances up to 5 km. The instrument measures two target ranges (only one of which is displayed under the operator's control) and records the total number of echoes within the beam. Angular measurements in azimuth and elevation are also displayed. Direct data output to a fire control computer is possible.

Laser Rangefinder—'LAR'

Netherlands

LAR is a two man portable 34 kg equipment which measures ranges up to 10 km. The laser source is neodynium glass. Power at 24 volt d.c. is supplied by a rechargeable nickel-cadmium battery pack and is sufficient for 150 measurements. A range finding pulse, when reflected from the target is processed into an electric signal resulting in a digital display or data output direct to a fire control computer. The pulse rate is 6 per minute. Azimuth and elevation angles are also measured, and with range information can be stored for a short time or displayed immediately. Ranges of two targets (separated by a minimum of 30 m) which fall within the beam can be measured, and both can be stored. Accuracy to 5 m is claimed.

Simrad laser rangefinder LP 3

Norway

The LP 3 is tripod mounted, man portable equipment weighing only 11 kg. The laser source is neodymium-doped glass and the pulse rate is 12 per minute. Maximum range is 20,000 m (minimum 200 m), range accuracy is claimed as 5 m, and range discrimination between echoes is about 30 m. Angles of azimuth and elevation can be measured to a resolution of one mil. Output is displayed on a digital indicator. A 7× magnification sighting telescope is fitted. The LP 3 can be fitted on to the British night observation device for use in conjunction with it.

Simrad mounted on British night observation device NOD-B

Laser rangefinder VAL 11105

Sweden

The VAL 11105 is available as a portable direct operating unit mounted on a tripod, or in a vehicle mounted periscope configuration. It operates to nearly 15 km. A neodymium glass laser source is used, powered by 12 rechargeable nickel-cadmium cells which are sufficient for 250 operations. There is a digital display of range and angles of azimuth and elevation, and a direct data output to fire control computers. The VAL 11105 is made by L. M. Ericsson, who also make the Naval VAL 10102. This heavy equipment (33 kg) has a Ruby laser source and has a range of 30 km with a claimed accuracy to 10 m. It has been adapted experimentally for land use.

AN/GVS-3

United States of America

The AN/GVS-3 weighs only 14 kg. It has a maximum range of 10 km, and a minimum range of 250 m. Range measurement is accurate to 5 m. There is provision for angular measurement in azimuth.

TM 12

France

This is another lightweight laser (15 kg). Its measurement range is between 300 m and 10 km with an accuracy of 5 m range, 2 mils in azimuth and 1 mil in elevation.

Night Observation Devices

An artillery observer must be able to cover his zone of observation by day and night. The night observation devices which are currently being developed are image intensifiers. These make possible identification of moving targets out to about 2,000 m with even greater range in moonlight, but the snag is that any white light on the battlefield renders them temporarily useless, although in-built automatic brightness control is overcoming this problem. The next generation of night observation devices is likely to rely on thermal imaging, which will also have the advantage of being able to see through battlefield smoke and mist. Many different varieties of device will soon be available, although few are yet in service. They will all tend to be bulky due to the need for a large viewing lens. The usefulness of the passive night observation device is greatly increased if it can be used in conjunction with a laser range finder.

Short Range Surveillance Radar

A short range radar is useful to the artillery observer by night or in poor visibility. It is able to scan out to the limit of line of sight and to distinguish between moving vehicles and men. Its effective arc of surveillance is about 60° with detection ranges being possible out to 10 km. The radar will produce bearing and distance to the target, and it is then a simple matter for the observer to find its co-ordinates. Operator fatigue is a problem, but this can be alleviated to some extent by incorporating an audio alarm signal.

Examples of short range surveillance radars are: Radar No 14 GS Mk 1 (ZB 298) (United Kingdom), Rasura DR-PT-2A (France), Rasit 72 A1 (Rapier) (France).

3. SURVEY INSTRUMENTS

Accurate survey of the locations of guns, observers and targets is an essential step towards achieving first salvo accuracy—they must all be on the same survey grid. Up to now this has been achieved only after laborious and time-consuming instrumental measurement and manual computation by artillery surveyors. The introduction of an inertial survey instrument—the Position and Azimuth Determining System (PADS)—will transform this situation. PADS is a distance and azimuth gyroscopic instrument, which in the British version shortly to be introduced, is basically the same as that developed for the Harrier fighter/ground attack aircraft. It will give an instant digital display of a vehicle's coordinates with an accuracy of a few metres, and it is intended to be issued to all observers, gun position command posts and target acquisition devices. Its limitations are that its survey data base has to be up-dated frequently, and every 10 miles or so, against the data of previously surveyed bearing pickets. The United States and France are known to be developing survey instruments which are similar to the British PADS.

PADS display in Landrover

Even with a wide distribution of PADS instruments, there will still be a requirement for skilled artillery surveyors to provide the data for bearing pickets (i.e. a post, or picket, clearly marked with its co-ordinates and the bearing to various natural reference objects to enable an azimuth angle measuring instrument—theodolite or director—to be set up) and for those who require survey but do not have PADS throughout the area of operations. Above all they require very accurate distance measuring instruments and these are normally commercial instruments adapted for military use. A British example is given below.

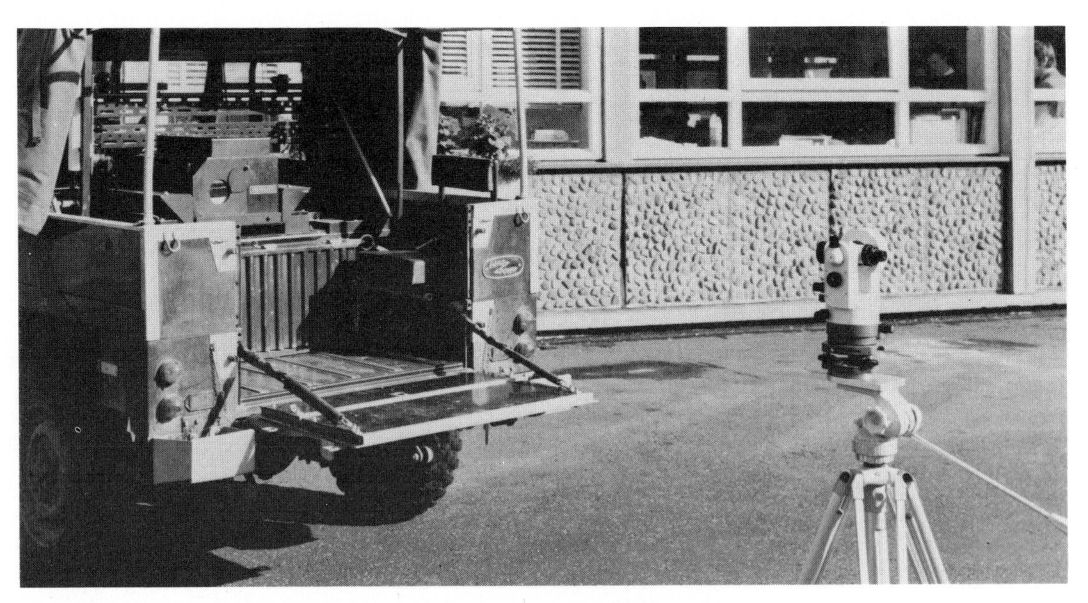

Landrover mounted PADS system

Tellurometer MRA 5

United Kingdom

The equipment comprises a control unit, an antenna unit and various cables and connectors. It can all be carried in a landrover. Two instruments are required to make a measurement, a master and a remote set. The accuracies obtained are better than 1 in 100,000. Microwaves are transmitted to a reflector at an identical out-station, and the time interval between transmission and reception provides an extremely accurate measure of the distance between the two stations. The result in metres is displayed instanteously at the master station.

The Tellurometer set up for use

4. METEOROLOGICAL INSTRUMENTS

Information about meteorological variations essential to compute or 'predict' variations in the trajectory due to abnormal wind, temperature and barometric pressure must be regularly received by the artillery command posts. It must be gathered at frequent intervals by the meteorological station and disseminated quickly. The usual method is to measure the rate of ascent and wind blown track of a balloon. Radar tracking is used and the resulting readings are processed by computer into several 'standard' measurements, each tailored for a specific type of equipment, i.e. guns, missiles, sound-ranging, drone systems, etc. The message can be disseminated by voice, teleprinter, or as a direct data transmission to fire control computers. Examples of some meteorological systems in service are given below.

Artillery Meteorological System—'AMETS'

United Kingdom

AMETS is a self-contained, mobile, upper air meteorological station, designed to provide up-to-date information rapidly. Many of the component parts of AMETS are common with FACE, the British Artillery fire control computer—an advantage in production and maintenance.

The AMETS system is completely mobile and largely automatic in operation. A single AMETS will normally provide the meteorological data required in a divisional area.

The balloon-carried radiosonde needs no pre-flight calibrations and is automatically tracked by a modified Plessey WF 3 radar, after initial visual tracking.

Information is processed by the Marconi 920B computer and can be printed out for voice transmission, or produced as a punch tape suitable for direct data transmission to FACE. A meteorological message covering up to an altitude of 20,000 m can be produced hourly. A message for any given height can be produced within three minutes of the radiosonde passing that height. Information messages can be produced for guns, drones, sound-ranging, and fall-out calculations for chemical and nuclear defence.

AMETS lightweight system deployed in the field

AN/UMQ-7

United States of America

The AN/UMQ-7 is a mobile equipment system which has its own 45 kilowatt generator, hydrogen generator tracking radar and computer. Accuracy of tracking up to 160 km is 16 m for range and 0·05 m in elevation. It produces all standard meteorological messages.

The observation and computation of data and the production of standard messages is fully automatic.

Sirocco

France

The Sirocco system is currently undergoing trials in France. It comprises a meteorological balloon tracking radar (QR-MX-2) and a telemetry receiver providing wind speed and direction and air temperature at a series of standard altitudes. It passes standard messages on a teleprinter or over radio links to field artillery units. This may either be fed directly into the Atila fire control computer, or used for manual computation of firing data. Maximum range of the radar is in excess of 130 km with a tracking accuracy of 5 m for range and better than 1 mil angular accuracy. A wind speed measurement accuracy of better than 1 kt is claimed.

5. MUZZLE VELOCITY MEASUREMENT

Before the advent of radar, muzzle velocity (MV) was calculated from a series of test firings to observe the fall of shot on a measured range, by firing through screens, photographing the projectile as it left the muzzle and by extrapolation from wear measurements made in the bore or by the number of rounds fired. The need for greater accuracy of fire demands measurement on the operational site; repeated at intervals to establish the performance of a gun as its wears. Variation in muzzle velocity is caused by small manufacturing variations in the weight of projectile and propellant, and by propellant temperature as well as by the gradual wear of the bore. The variations are significant in their effect on the fall of shot and must therefore be considered in the calculation of gun-sight data. A number of velocity measuring devices are now available which can be used at the gun position to monitor changes in muzzle velocity. Some feed information directly to the command post computer. They all work on the Doppler principle and consist of a radar head and small computer which translates the shift in frequency between the outgoing radar pulse echo from the shell into velocity.

In order to avoid geometrical errors, the radar head must be close to the muzzle. This exposes the mechanism to the shock of firing, so that the equipment must be robust. At the moment of leaving the barrel the propellant gases create an ionizing effect so that measurement must be made after the shell has travelled some distance from the muzzle. To achieve the highest accuracy several measurements are made and the results compared electronically to extrapolate the muzzle velocity.

DR 513 ballistic radar

Denmark

The remarkable feature of this doppler radar is its rugged construction and high shock and vibration resistance which enable it to be placed very close to, or even on, the gun being examined. This ensures the least possibility of geometrical error in the result. It is a compact and self-contained equipment suitable for both field and range use. A test unit is incorporated in the design. Over 14 countries have been supplied with the DR 513.

Velocities from 50 to 2,000 m/sec can be measured to an accuracy of 0·1 per cent. There is a visual readout of muzzle velocity on a tape which can be fed to a computer and is sufficient to extrapolate the shell's performance at any point on the trajectory.

The horn antenna incorporates both transmitter and receiver and can be placed on any non-recoiling part of the gun or on a tripod. The data box is remoted from the antenna. Measurements are made automatically at 30 and 40 m from the muzzle.

Note transmitter to left of wheels—antenna beyond

DR 810 ballistic radar

Denmark

The DR 810 is designed specifically for field use. Like the DR 513 it can be mounted on or very near the gun being examined. The antenna weighs 13 kg, and the data box 22 kg. 24 volt d.c., or 220 volt a.c. electrical supply can be used without recourse to additional equipment.

The DR 810 measures velocities from 50 to 1,750 m/sec. The projectile velocity is measured at several points along the trajectory and the muzzle velocity calculated and displayed immediately. The first measurement point can be varied to avoid the disturbing effects of muzzle flash. The Nixie tube display in metres/second can be of one firing, or an average of up to eight previous firings.

There are selectors for the velocity band, one of three first measuring points (30, 60 or 120 m), and the offset of the antenna. Compensation for parallax is automatic.

Isidor muzzle velocity meter

Sweden

The Isidor meter is intended for field use. It consists of a light solid state antenna (four kilogrammes) and a data unit. It can be mounted on the gun being examined.

The Isidor measures velocities from 175 to 1,150 m/sec. The read-out can be the measurement of one firing, or an average of several shots. The equipment operates to a range of 2,000 times the calibre of the weapon being measured. Accuracy of 0·5 per cent is claimed.

The direct read-out is expressed as the difference between the nominal performance of the gun which is pre-set by the operator, and the actual velocity measured.

Projectile velocity measuring radar—'Pacer'

United Kingdom

Pacer is designed for use in the field. It is small and operates from a 24 volt d.c. or mains voltage a.c. electrical supply. Unlike the Scandinavian systems, Pacer is not intended for semi-permanent attachment to the gun, but is deployed (by one man) when necessary. It weighs 130 kg.

Pacer measures velocities between 200 and 1,600 m/sec to an accuracy of 0·3 m/sec. It can measure down to 100 m/sec with reduced accuracy. There is a built-in automatic test facility.

The operator must select the anticipated muzzle velocity to within 20 m/sec: the radar is then automatically activated by the muzzle flash, and after the measurement is made, automatically switches off: the result of the measurement is shown on a Nixie tube display. The only other actions required by the operator is the initial setting up of the radar head within 2 m of the barrel between muzzle and breech, and aiming it along the line of fire.

Pacer—transmitter (*on left*) and computer/display unit.

Projectile velocity measuring equipment (PVME)

United Kingdom

PVME is complementary to Pacer, and is designed for experimental, research and proofing uses. It is both more complex and more accurate. Both PVME and Pacer are manufactured by Ferranti Ltd. and a combination of parts of the two equipments is possible so that the rugged and shock resistant Pacer antenna can operate with the PVME data handling equipment.

An accuracy of 1/10,000 is claimed. All measurements are duplicated by two transceivers operating in different radar bands. The points of measurement on the trajectory are variable and controlled by the operator. There is an oscilloscope display of velocity against time and a stored display of the timings of the projectile at the three measuring points. A tape for subsequent print-out or computer input can be provided.

6. FIRE CONTROL COMPUTERS

The development of the computer for field use has been the most important technical innovation in gunnery since World War II. In most cases the military computer has been developed from a commercial design, and specially adapted to cope with the tough battlefield environment.

The field artillery computer is used to resolve firing data for the gun sights from the co-ordinates of gun position and target to include allowance for all known ballistic variables. The computer programme includes all known performance facts for the type (or types) of gun being served. Variables such as individual gun muzzle velocities, changes in temperature and meteorological conditions can be fed into the computer manually, by tape or by direct data link from a suitable instrument.

Manual input requires a keyboard, often associated with a matrix of light symbols which indicate which part of the computer programme or memory is being used.

The computer is a flexible instrument which can be programmed for any type of gun or for many other artillery problems: survey meteorological calculations, sound-ranging and logistic problems can all be automated. Most small field service computers need to be re-programmed before tackling each type of task; only the largest can handle all calculations using one programme. Programmes are usually fed into the computer by tape—a simple process which can be performed in the field by the operator: in some cases the matrix must also be changed.

Computers are expensive, but they are fast and highly accurate: they have in-built test and error detection facilities, require less training for the operator and do not suffer from battle fatigue or human error.

Falke computer

Federal Republic of Germany

Falke is based on the Telefunken TR-8 family of digital computers and is known as the TR-84M. It is one of a family of military computers. It produces gun-sight data, calculates meteorological data, resolves sound-ranging and flash-spotting observations into hostile battery co-ordinates and assists fire control.

Field artillery computer—Saab ACE-380

Sweden

The ACE-380 is a field artillery computer weighing only 50 kg. It is based on the Datasaab D5-30 and is designed to be used in a battery command post. It calculates firing data from one gun position, taking into account the existing ballistic variables. Meteorological corrections are added from data processed by the computer itself based on measurements at any 30 levels of atmosphere. There is an automatic error detection device.

After manual entry of gun position information and meteorological data, target data can be fed into the computer automatically or semi-automatically by the observer, or manually by the operator. A keyboard is used which has six function areas: battery co-ordinates, target data, change of target, fire adjustment and function controls such as 'compute', 'test' and 'store'. Muzzle velocities are automatically entered from measuring devices on the guns.

Keyboard and display panel

There is a visual display of firing data in the command post and a direct data link to the gun sights.

A ferrite core memory stores up to 16,000 words. This is sufficient to handle 500 targets together with 20 target records corrected for current conditions. Stored data can be recalled, displayed, up-dated and restored, as required.

Field artillery fire control equipment—9FA 101

Sweden

The 9FA 101 is intended to be a simple battery command post computer. The system can be extended to include automatic input by forward observers with a data link direct to the guns. The basic computer weighs 31 kg.

The main function is the production of gun sight data from target and gun position co-ordinates. 9FA 101 also produces fuze settings and gun corrections from observations of the fall of shot.

Target co-ordinates are fed to the computer by keyboard, while other relevant data is set on multi-position switches. A direct radio link input device for forward observers called HADAR is under development.

All firing data is displayed numerically, but can be augmented by repeater stations at each gun. The repeater units incorporate speech facilities for the passage of supplementary orders.

Field artillery computer—'FACE'

United Kingdom

FACE is a digital computer based on the commercial Marconi-Elliott 920B. It has been in service with the British Army since 1969. The system is rugged and suitable for rough combat zone use. It runs off a 24 volt d.c. electrical supply, and can be mounted in a Landrover. It has been sold to three other countries.

FACE can be programmed to produce gun or a free flight rocket firing data, or to solve survey problems. It is also incorporated into AMETS. In its gun control role it can handle calculations for three battery positions, each of eight guns. Only one type of equipment can be handled by each programme. The programme can be changed by the operator in the field, and when a new display matrix is required, this too can be changed

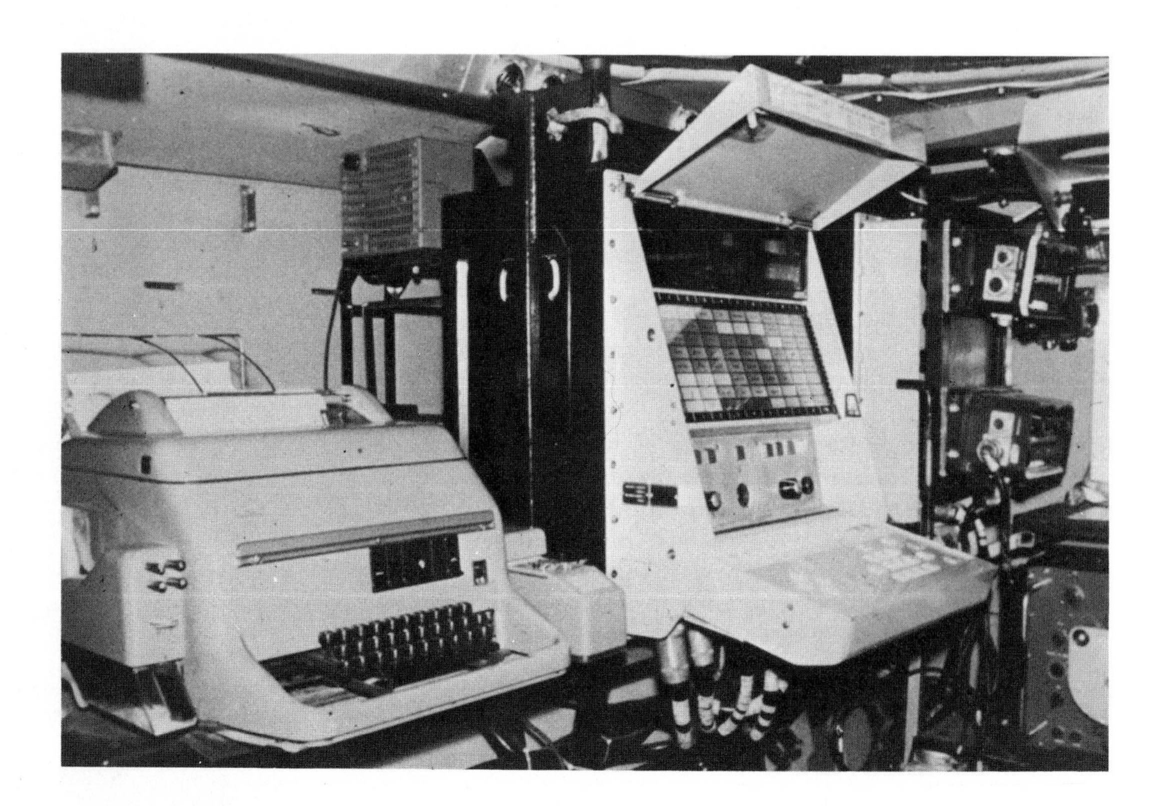

FACE layout in FV 432 command post vehicle

by the operator. The programme includes the drill sequences, so that after each input the next piece of information is requested by an illuminated panel in the display matrix.

The required programme including the ballistic data of an individual weapon, is entered via a sealed cassette tape. Target data and variable factors such as individual gun muzzle velocities are fed into the memory by the operator using a key console. Meteorological data is inserted by punched tape (available from the teleprinter) or via the key console.

Data is shown on a Nixie tube visual display. A printed record, or punch tape read-out is possible using the built-in teleprinter. In the gun control role, firing data for each gun is displayed in sequence. Alternatively individual gun sight data can be transmitted to each gun by the Artillery Weapon Data Transmission System (AWDATS) which operates by FM radio frequency signal over existing radio or line equipment.

Besides the programme, initial battery position, individual gun data, and meteorological data, the FACE memory will hold 40 recorded targets, of which ten are automatically and permanently updated for current meteorological conditions.

Employment Australia, Canada, Egypt, UK

AWDATS display in turret of British 105 mm SP Abbot

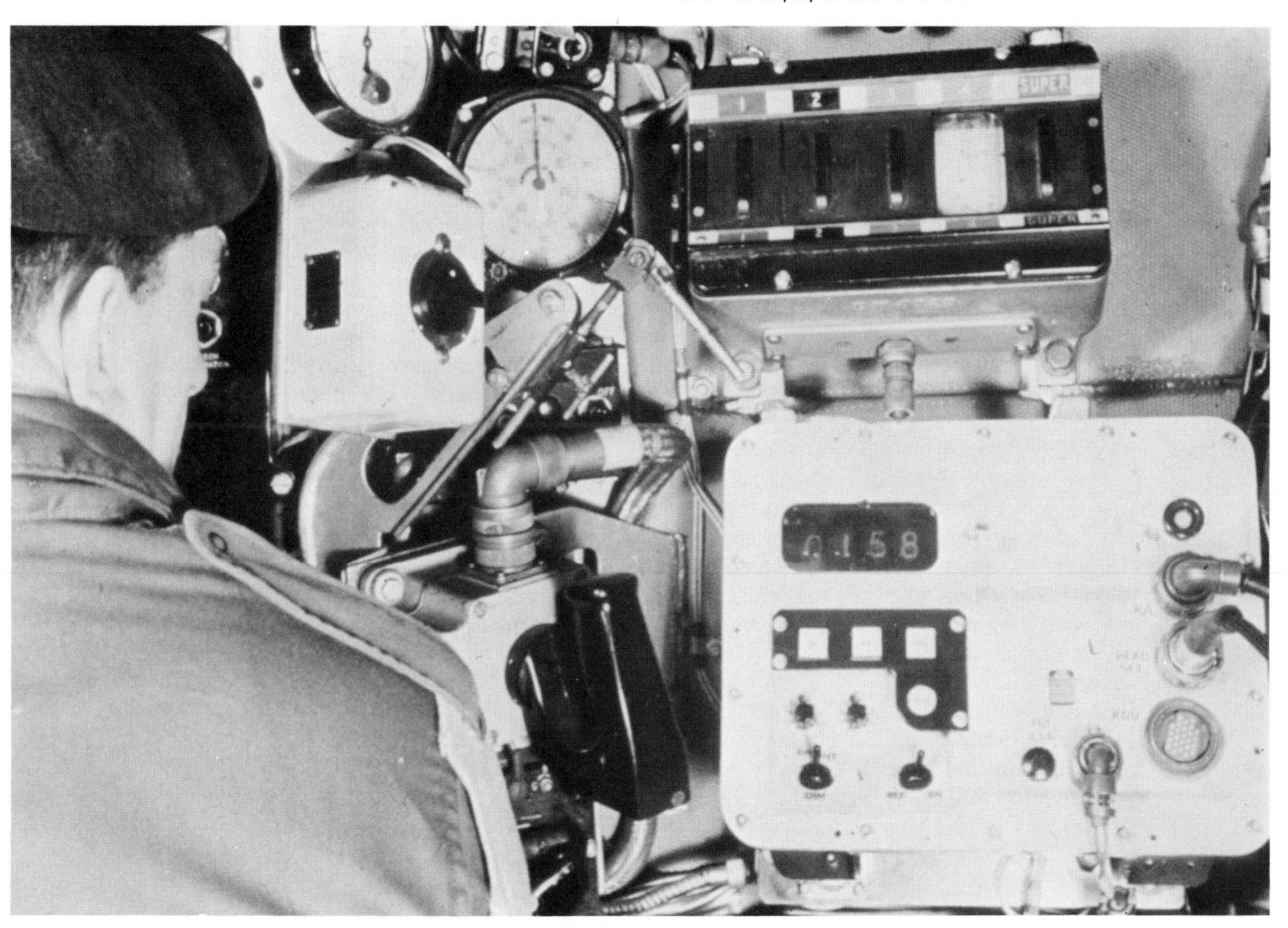

Field artillery digital automatic computer system—'FADAC'

United States of America

FADAC is a first generation field artillery computer system which processes a wide variety of data for tube artillery, guided missiles and free flight rockets. It consists of a computer, test set and programme input unit: a teletypewriter can be coupled to it, and this is used for survey, meteorological data processing and Lance missile fire control procedures. For normal fire control a digital display of data is used.

FADAC can be used to compute gun-sight data from target co-ordinates for up to five battery positions. It can also resolve survey problems, reduce meteorological information into gun corrections and assist in producing timed fire programmes. It is used to process data from sound-ranging equipment.

Programmes for each different task are kept on tape and fed into the computer through a photo-electric type reader—the 'programme input unit'. No other special equipment is needed so that FADAC can change roles in the field. Individual problems are fed in via a key console. For each role a different input matrix is needed to interpret the keyboard, but the operator can change the matrix himself. Output is by numerical display. The programme for use with guns allows the memory to store data for two calibres or types of guns (rocket and missile programmes are for individual equipments). Data for firing battery positions is stored; the operator selects the battery for which data is required. FADAC is no longer in production and it will be replaced in the US Army by a new battery computer.

Tactical fire direction system 'TACFIRE'

United States of America

The TACFIRE system is designed around the L 3050M computer. It is deployed at both battalion and divisional Fire Direction Centres. It handles the entire fire control process from the direct input of target data by the observer to the display of firing data in the battery command post. It stores gun and ammunition availability states, and the tactical situation expressed as the disposition of enemy and friendly troops. It can handle survey and meteorological calculations.

There are three methods of entry: manual or by fixed or variable format entry devices. Forward observers are equipped with a fixed format message entry device by which target information is codified using 25 key switches. The device transmits this information in digital form over existing radio or wire link communication systems in a 1·3 sec burst transmission. The device is also used to send corrections to adjust the fall of shot. The variable format message entry device is used by survey parties, battery command posts, meteorological observers and other 'out stations' to enter information relevant to their tasks. Direct entry is by keyboard.

At the computer console output is either visual or printed. When visual display on a cathode ray tube is used, a print-out record can also be kept. The usual fire control display which results from an observer's input of target data is in the form of a plan for target engagement based on available fire units which are in range, ammunition states and the scale of response suggested by the type of target. This plan can be accepted or modified before being routed to the relevant battery command posts, where it is displayed numerically as gun sight data.

The main memory unit is part of the divisional Fire Direction Centre equipment. It consists of a Random Access Memory which holds 400,000 words of information and a Bulk Media Memory Unit with a 300,000 word capacity in computer programmes. There is also a Memory Loading Unit by which information stored on tape can be recalled: the tape is used for initial programming and carries 300,000 words.

Development
ATILA (France) will be a comprehensive fire control system on the lines of TACFIRE, but it will go further in that it will include ADP at the battery level. MILIPAC (Canada: see below), TOS (USA) and ARCOS (UK private venture) are light-weight second generation battery level computers.

Field Artillery Computer 'ODIN'

Norway

This is a field artillery computing system in use with the Norwegian Army since 1972. It consists of a general purpose computer in battery and batallion command posts linked with SIMRAD laser rangefinders in the observation posts, a meteorological information system, doppler muzzle velocity measuring equipment and standard survey equipment.

The basic computer can handle firing data for 105 mm and 155 mm howitzers, survey problems such as change of grid, crest clearance, bearing and distance calculations for up to 100 targets and nine gun locations, a safety-zone programme for practice firing, calibration data for up to 18 guns and eight ammunition options, and the standard NATO meteorological message.

Field Artillery Computer 'DAVID'

Israel

This is a lightweight portable fire data computer in service with the Israeli Army since 1977. It allows computation of data for up to six sub-units, simultaneous conduct of two fire missions, and storage of data on up to 28 targets. It can also deal with survey, meteorological and calibration tasks. Output units permit transmission of data to individual guns, with separate display units for gun commander and gunlayers.

Field Artillery Computer 'MILIPAC'

Canada

One of the objections voiced against many contemporary field artillery data computers is that they are tailored too specifically to one particular weapon or to one particular form of military organisation; in addition, they tend to be extremely expensive and demand vehicles of their own.

To answer this and other complaints, 'MILIPAC' has been developed by Control Data Canada as a versatile and portable system. By use of plug-in modules the basic computer can deal with ballistic calculations for varieties of weapons in the 105 mm–155 mm bracket, and it can also perform a number of survey tasks and meteorological computations. Future planned additions include modules for accepting data from muzzle velocity chronographs, sound-ranging, and moving target prediction.

At present, prototypes are undergoing evaluation and test with the Canadian and Spanish Armies.

MILIPAC Computer

Battlefield Artillery Target Engagement System—'BATES'

United Kingdom

BATES will be a comprehensive command, control and communications system embracing all the artillery resources, command posts and artillery staff cells in a corps. Its functions are defined as *resource management*, *technical artillery computations*, *routine tasks*, *the provision of the primary artillery control network* and *assistance in making engagement decisions*.

Its elements are applications, processors and stores, distributed throughout the system, where all relevant artillery data is stored and continually updated, and which also provides gun-data for fire missions on demand (it will supersede FACE), input points (data entry devices DED) located with FOOs, artillery commanders and locating units where information and fire missions can be tapped in, visual display units (VDUs) supplying information on every subject relevant to a mission and plotter, and a printer for producing messages, orders and artillery overlays. Gun-data for fire missions are displayed at each gun on a gun display unit (GDU).

The communications system is digital and the system works in real time. There are sufficient elements and different pathways to enable BATES to function if subjected to battle damage, and the whole system is backed up by the ordinary voice-radio and cable network.

BATES is multi-functional, but the heart of the system is that it enables an artillery commander to retrieve at any moment the status of his weapon systems in terms of readiness, ammunition state, range, whether engaged or not, etc, and, when presented with a fire mission or a demand for a fire-plan, to be offered for approval options of engagement plans in terms of different types of batteries and ammunition.

Hitherto the Royal Artillery has clung to two basic principles for the application of fire: to rely completely upon the decisions of junior officers in contact with the situation in the battle zone, and to turn every gun in range on to large or dangerous targets. BATES is a revolutionary system, imposing a completely different approach, and may well prove to be a step forward as decisive as the introduction of indirect and predicted fire.

Employment UK (design stage)

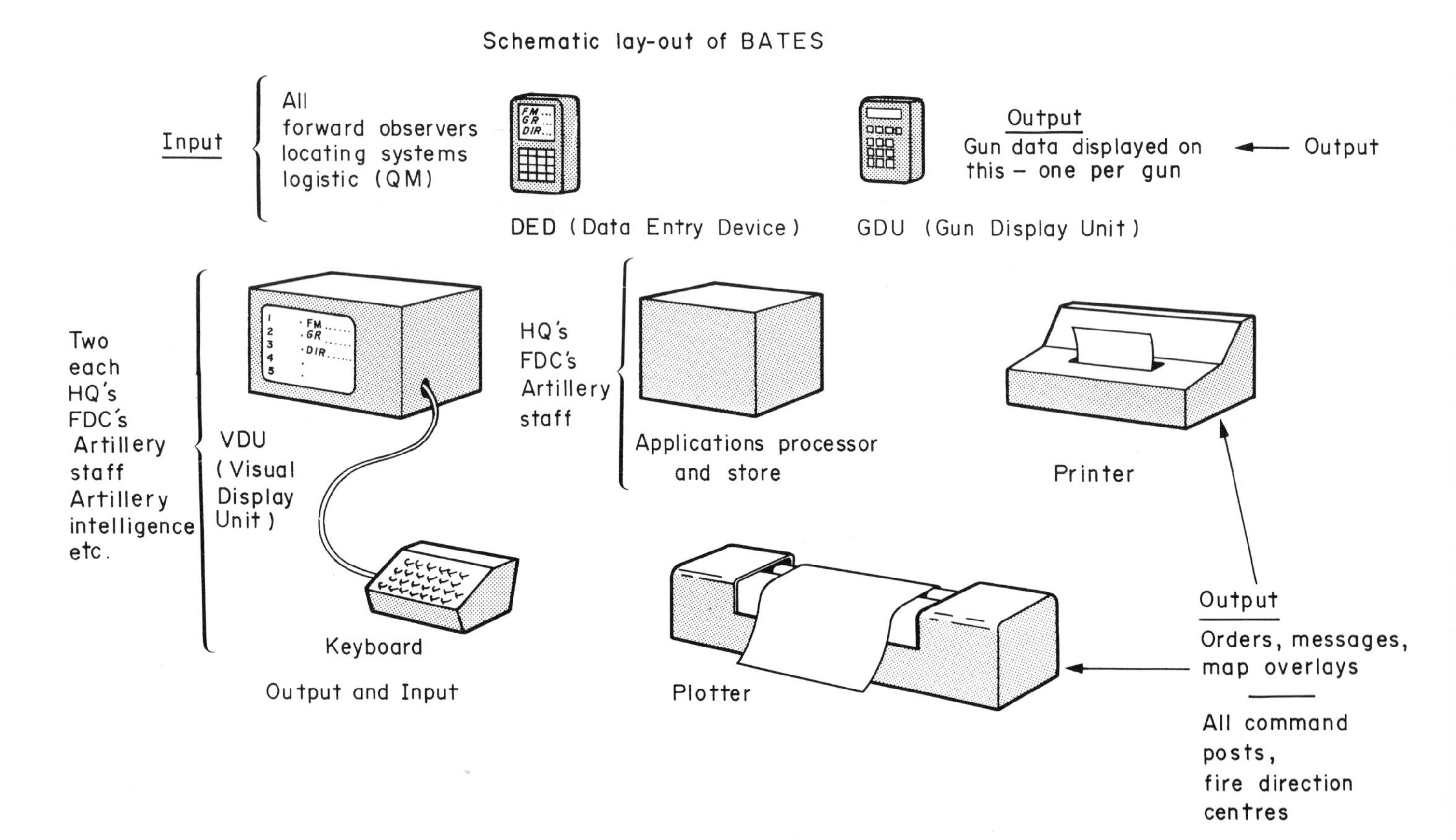

VIII. Anti-tank Guns and Guided Weapons

The dominating position occupied by the tank in modern land warfare has naturally led to the development of a whole spectrum of anti-tank weapons. At one end there is the tank itself which has developed into a gun-platform for a gun primarily designed to defeat other tanks, and at the other there are a variety of hand-held infantry weapons. There are no clear-cut divisions, but there are, broadly, four classes of weapon.

First there is the true 'gun', a long, powerful, heavy, high-velocity weapon of great accuracy relying on the kinetic energy of the shot to achieve its effect. This is not only the armament of the tank, but also of the tank's close relative, the self-propelled anti-tank gun. The difference between the two lies in the fact that the SP anti-tank gun is primarily a defensive weapon as opposed to an armoured fighting vehicle suited to all phases of combat, and it is accordingly more lightly armoured and not necessarily equipped with a revolving turret. Tanks are so stoutly armoured today that guns of a size able to generate the kinetic energy required to penetrate it using a solid shot or dart are more conveniently carried on a tracked chassis. Towed guns of the size of the 17 pdr and the 100 mm are slow to manhandle in and out of action. The SP gun in addition has obvious advantages as its armoured protection and mobility enables it to take part in a rapidly moving tank battle.

The towed equipments, fitted like the tanks with a high velocity gun, may either be used primarily in the anti-tank role or may also be fitted with a dial sight and clinometer for indirect fire using HE shell. The Russians, who have always favoured guns, consider that their indirect fire guns also have an important secondary role in the whole anti-tank defence lay-out. As regards design, the requirement is not so much a high sustained rate of fire, but the ability to engage successive targets very quickly, so the tendency is towards semi-automatic sliding breech-blocks and fixed ammunition. As the gun has to be deployed well forward and concealed the carriage silhouette is kept low and a generous shield is provided. Some Russian towed guns are fitted with auxiliary propulsion units.

The weight of the larger towed guns reduced their mobility and accordingly the recoilless rifle firing a special armour-piercing HE shell was developed. In the recoilless system a portion of the gas generated by igniting the propellant is vented through the breech to the rear to counteract the recoil. This enables the whole buffer/recuperator system and the trail to be dispensed with and so saves a great deal of weight. The muzzle velocity is low which also makes for a thin walled and lighter piece, but the resultant curved trajectory demands some form of on-carriage range-finding device. The great disadvantage of the recoilless system is that there is a large back-blast which on sandy or dry ground gives away the position of the gun.

The normal sighting system used on anti-tank guns is the rocking bar with the telescope. Range is either judged by eye, or determined by a spotting rifle firing a trace or an indicator round to test the range or, coming increasingly into use, a laser range-finder. The lead to allow for the movement of the target is usually estimated (or guessed) and is applied either by off-setting the telescope, or more usually by aiming off, using graticules which appear in the telescope's field of view. Anti-tank guns require considerable skill in

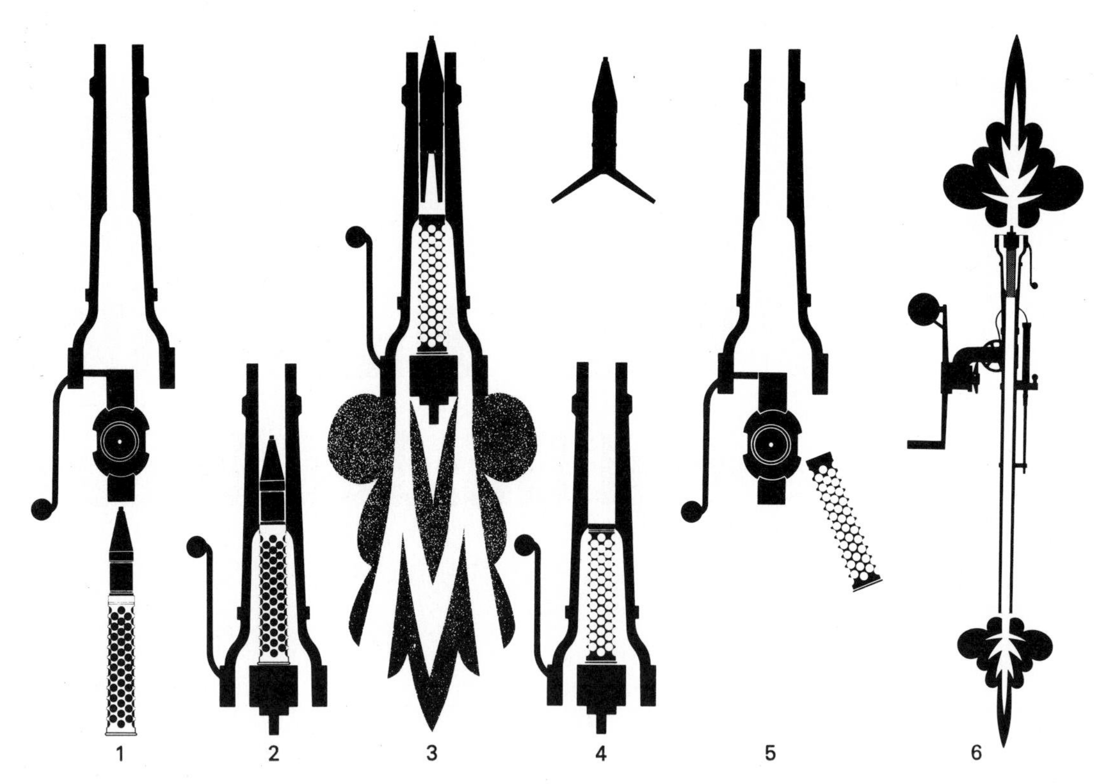

Figure IX: Principles of the recoilless rifle. *From L. to R.*: 1, 2, round with perforated shell case is loaded. 3. Firing, gases vented to rear. 4 and 5, projectile leaves, shell case ejected. 6, general idea of system, muzzle at bottom of page

deployment, and brave, well-disciplined, well-trained crews, as once a tank battle starts they attract every form of counter-fire. Only a really skilled layer can be expected to obtain first time hits, which are what count, for the expectation of survival of a gun which has exposed itself by firing and has missed are short. Accordingly attention has turned to the guided missile, which first became prominent in the Vietnam war and later in the Arab–Israeli war of 1973, as its launcher is easy to conceal and the probability of a first-time hit at any point along its effective trajectory high.

Like mortars, anti-tank weapons of all kinds are used by all three arms, but only the long-range, anti-tank guided weapons (LR-ATGW) are considered here, as they can be regarded as a crew-served and, in the case of the British Swingfire, can be fired indirect from behind cover and so are 'artillery'. With the exception of the Soviet Swatter they are all wire-guided; receiving signals through a thin filament which pays out behind during flight. Guidance is of two kinds. 'Command to the Line of Sight' requires the operator to steer the missile on to the target. In a 'Semi-Automatic to the Line of Sight' all the operator has to do is to keep his aiming telescope on the target and its movements are translated into corrections signalled to the missile via the wire.

Missiles are easily deployed and concealed, are versatile (man-handled, vehicle mounted, AFV mounted, or airborne in helicopters); as said, the first time hit probability and lethality are very high, accuracy is not degraded with range, crew-training is short and easy and the exchange ratio in terms of cost between tank and missile is greatly in favour of the missile. There are, as always, compensating disadvantages. There is the question of missile re-supply and the carriage of first-line holdings, as a missile is fragile and bulky compared with an artillery round, the operators of the direct fire variety are very exposed to retaliatory and suppressive fire, and the time of flight is long and the time between engagements also long in consequence. To offset these a large number of missiles and operators must be deployed forward, which was the way the Egyptians obtained their first, surprise effect in Sinai in 1973. The optically controlled wire-guided system has its limitations, but an active or semi-active homing head and a radio link is open to 'spoofing' and electronic counter-measures. Future development is likely to be aimed at the fully automatic weapon which locks on target and homes after launch which can engage out to the limit of target acquisition by day or night and enable a succession of targets to be engaged at intervals of seconds. This of course will inevitably raise complexity and cost.

It must be understood that the above is simply a brief account of the situation at the time of going to press, but nothing in the state of design is permanent. As is always the case in the history of weapons, whenever a predominant weapon appears, there is intense research for a counter-weapon, which in turn generates improvements to defeat the counter-weapon. For instance, for forty years the tank has been progressively more heavily armoured, until sheer weight brought this response to a halt. Now a totally new form of armour, much lighter in relative terms of resistance to penetration and thickness has appeared—the British invention of 'Chobham' armour, whose composition is secret. The tanks themselves, considered by many to be the most effective anti-tank weapon, are progressively being armed with ever larger guns with higher muzzle velocities and better projectiles. Some years back it looked as if the combination of the helicopter and the guided missile threatened the very existence of the tank, but this has been met by the appearance of both light anti-aircraft guns with a high cyclic rate of fire and SAMs mounted on armoured tracked vehicles deployed forward. At the same time aircraft bombs and indirect fire artillery projectiles with good anti-tank performance against even the heaviest modern tanks are under development or in service. The only limit to this progress is cost—which of course applies to the whole field of weapon development.

6 pounder anti-tank gun

United Kingdom

Calibre	57 mm
Barrel length	50 calibres
Muzzle brake	Single baffle
Trail type	Split
Weight	1,244 kg
Ammunition	APDS
Armour penetration	146 mm at 1,000 m (APDS)
Rate of fire	20 rds/min
Muzzle velocity	1,235 m/sec (APDS)
Maximum range	5,030 m
Maximum effective range	500 m
Elevation limits	−5° + 15°
Traverse on carriage	90°
Detachment	6

The 6 pounder was developed at the beginning of World War II to replace the 2 pounder anti-tank gun. Design work started in 1938 and the gun entered service in 1942.

The monobloc barrel has a distinctive spherical muzzle brake and a vertical sliding wedge type breech. The recoil mechanism is mounted below the barrel. The shield is angled back into extensions over the wheels and has a hinged flap to give line of sight to the optical telescopic sights.

The split trail of box section members is of conventional design, but opens wide to allow excellent traverse.

The weapon was produced in America as the 57 mm M 1. It was used in various mounts including tanks and as an SP anti-tank gun on an armoured truck (known as Deacon). It is long since obsolete in the British and American armies.

Employment			
	Bangladesh	Egypt	Pakistan
	Brazil	India	Spain
	Burma	Israel	Thailand
	Cameroons	Malaysia	

6 pounder Mark IV anti-tank gun

Type 56 recoilless rifle

China

Calibre	75 mm
Barrel length	28 calibres
Muzzle brake	Nil
Trail type	Single support strut
Weight	86 kg
Ammunition	HEAT 80 mm HE
Armour penetration	80 mm (HEAT)
Rate of fire	10 rds/min
Muzzle velocity	—
Maximum range	6,600 m
Maximum effective range	500 m
Elevation limits	—
Traverse on carriage	360°
Detachment	4

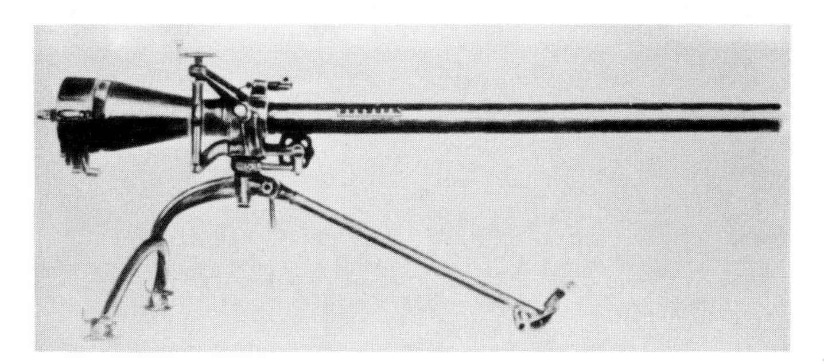

75 mm Type 56

The Type 56 recoilless rifle is a weapon developed by the Chinese communists from United States' equipment. The original was the M 20 recoilless rifle, and only the general form of the tube and breech remain. The mechanism of the breech is, however, considerably altered. Whilst retaining the conventional long, clean barrel and curved conical breech, the Type 56 is mounted on a distinctive curved tubular axle with a single forward tubular support ending in a transverse towing bar. The tube is connected to the carriage by a single ring clamp with the prominent elevating and traversing screws set at right angles to each other behind and in front of the ring clamp respectively. The Type 56 or Type 62 sight, mounted on the left above the clamp, permits either direct or indirect fire, using either HEAT or HE rounds, at a rate of up to 10 rounds/minute. It is possible to penetrate some 80 mm of armour. The weapon can be fired without removing the detachable solid-tyred wheels, but is, of course, more accurate and stable without them.

Employment China
Vietnam

---

82 mm recoilless rifle B-10

Union of Soviet Socialist Republics

Calibre	82 mm
Barrel length	20 calibres
Muzzle brake	Nil
Trail type	Tripod
Weight	85 kg
Ammunition	HEAT (3·6 kg) HE (3·2 kg)
Armour penetration	250 mm (HEAT)
Rate of fire	5–6 rds/min
Muzzle velocity	322 m/sec
Maximum range	4,400 m
Maximum effective range	500 m
Elevation limits	−2° + 35°
Traverse on carriage	360°
Detachment	4

This weapon dates from 1950 and derives from the RPG 82. It is no longer in service with the Soviet Army. Like the B-11 it is a smooth bore equipment, without muzzle brake and towed by its barrel. The projectile is spin stabilised and has the appearance of a mortar bomb. An HE round adds flexibility to the weapon's employment. The PB 02 sighting gear is clamped to the left of the barrel, while a pistol grip and trigger are to the right. The hinged breech opens horizontally.

This weapon is fired from an adjustable height tripod which gives 360° traverse. It travels on a two wheel carriage which can be removed or remain attached in action. A small castor wheel below the muzzle protects the barrel from damage while the weapon is being manhandled. It weighs only 72 kg in action and is thus within the carrying capacity of its detachment for manoeuvering. It can be mounted on BTR 50 APCs.

(illustrated overleaf)

Employment

Bulgaria	Hungary	Syria
China	North Korea	Vietnam
Egypt	Pakistan	
Germany (GDR)	Poland	

The 82 mm recoilless rifle B-10 deployed in Sinai

82 mm recoilless rifle M 60 (BO)

Yugoslavia

Calibre	82 mm
Barrel length	—
Muzzle brake	Nil
Trail type	Pole
Weight	122 kg
Ammunition	HEAT (7·2 kg) HE
Armour penetration	220 mm (HEAT)
Rate of fire	4–5 rds/min
Muzzle velocity	388 m/sec
Maximum range	4,500 m
Maximum effective range	500 m
Elevation limits	−20° + 35°
Traverse on carriage	360°
Detachment	2–5

This simple weapon consists of a barrel mounted on a two wheel axle with a single steadying leg.

The breech hinges downwards for loading. The PTDM 60 sights are on the left of the barrel, but a foresight is fitted to the right side to enable the detachment commander to indicate targets, and check the lay of the gun.

The standard ammunition is a HEAT round which is notably heavier than that of the Russian B-10 weapon. It is considered effective against static targets up to 1,000 m, but for moving targets 500 m is the practical limit. The armour penetration is sustained over an angle of incidence from 25° to 90°. The rate of fire is not very high for the type of equipment.

A range of 4,500 m for indirect fire is quoted, and the barrel has a generous elevation. However only anti-armour ammunition is provided, nor do there appear to be indirect sights, suggesting that the 'field artillery role' is only theoretical. The supporting leg folds under the breech for travelling, and the weapon is towed by its muzzle. There is no muzzle brake: two handles project horizontally from the towing eye mounting to assist in manhandling. At 122 kg it is an extremely light weapon and can be carried by its detachment over short distances.

Employment Yugoslavia

82 mm recoilless rifle M 60

85 mm anti-tank gun M 52

Czechoslovakia

Calibre	85 mm
Barrel length	56·2 calibres
Muzzle brake	Double baffle
Trail type	Split
Weight	2,095 kg
Ammunition	APHE (9·3 kg)
	HVAP (5 kg)
	HE (9·5 kg)
Armour penetration	123 mm at 1,000 m (APHE)
	107 mm at 1,000 m (HVAP)
Rate of fire	15 rds/min
Muzzle velocity	1,070 m/sec (HVAP)
	820 m/sec (APHE)
	805 m/sec (HE)
Maximum range	16,200 m
Maximum effective range	1,000 m
Elevation limits	−6° + 38°
Traverse on carriage	60°
Detachment	6

This conventional gun is designed to fire Russian 85 mm M 1945 ammunition: its longer barrel gives a marginally better performance than the Russian gun. The vertical sliding breech block lifts to open and is probably semi-automatic.

The rate of fire is high. The recoil system is mounted under the barrel.

The gun is primarily designed as an anti-tank weapon. However, it has an HE round and the high muzzle velocity gives an excellent indirect fire performance. The anti-armour penetration of both APHE and HVAP rounds is not good by modern standards.

The design of the carriage provides greater elevation and traverse than its Russian counterpart. The trail legs have fixed spades: the towing eye on the left leg folds inside the leg to clear the spade in action. A stowage box travels on the trail legs and is an important recognition feature. The shield, which has a wavy top, narrows at its base to give a triangular appearance.

Employment Austria
Czechoslovakia
Germany (GDR)

85 mm auxiliary propelled gun SD 44 (D-48)

Union of Soviet Socialist Republics

Calibre	85 mm
Barrel length	55 calibres
Muzzle brake	Double baffle
Trail type	Split
Weight	2,256 kg
Ammunition	HVAP (5·0 kg) APHE (9·3 kg) HE (9·5 kg)
Armour penetration	113 mm at 500 m (HVAP) 91 mm at 500 m (APHE)
Rate of fire	15 rds/min
Muzzle velocity	1,030 m/sec (HVAP) 792 m/sec (HE and APHE)
Maximum range	15,650 m
Maximum effective range	1,600 m
Elevation limits	−5° + 40°
Traverse on carriage	54°
Detachment	5–7

This is an ingenious SP version of the previous entry, the 85 mm M 1945. Performance is unchanged except in a reduction to the rate of fire, due to the mountings of the motor and ammunition bin which restrict the movement of the loaders. The detachment is reduced. The weight is obviously increased but only by a modest 525 kg.

The M 72 two cylinder engine is mounted on the trail legs and a castor wheel is fitted beneath the spades. The gun runs 'backwards', i.e. away from the line of fire as if it were being towed. The driver sits in front of the power pack behind a conventional steering wheel. The third wheel and steering wheel column fold in action, but the power pack remains on the left trail leg. An ammunition locker is carried on the right trail leg. There is provision for one passenger on the breech, and up to four passengers can be carried. The 22 bhp motor propels the gun up to 7·5 km/hr. Its fuel is carried in the hollow trail legs.

The gun can also be towed behind a vehicle and for this the third wheel is folded between the trail legs, where it is stowed in action.

The weapon is used by Soviet airborne forces.

Employment			
	Albania	Czechoslovakia	Poland
	Bulgaria	Germany (GDR)	Rumania
	Cuba	Hungary	USSR

The 85 mm SD 44 auxiliary propelled gun

90 mm anti-tank gun

Federal Republic of Germany

Calibre	90 mm
Barrel length	—
Muzzle brake	Double baffle
Trail type	Box
Weight	5,000 kg
Ammunition	HEAT HVAP Practice
Armour penetration	330 mm at 1,000 m
Rate of fire	12 rds/min
Muzzle velocity	1,181 m/sec
Maximum range	—
Maximum effective range	2,000 m
Elevation limits	−8° + 25°
Traverse on carriage	30°
Detachment	3

This is a self-propelled weapon, which fully exploits the principle of auxiliary propulsion, and which deploys into action as a conventional towed gun. It is in an advanced stage of development. It can be, and for long journeys always is, towed.

The 90 mm gun is taken from the Jagdpanzer Kanone, a fully SP anti-tank weapon. It carries a double baffle muzzle brake and is fitted with a semi-automatic breech. Loading is manual, and firing either electromagnetic or mechanical using a percussion initiation.

Infra-red night sights are fitted as standard equipment.

A box-type trail acts as the vehicle chassis. It carries the recoiling parts, the detachment of three men, and the motor. There is space for 51 rounds of ammunition. The rear wheels (of small diameter) are mounted in a subframe which folds up for firing. The action of folding the subframe emplaces a spade, and is hydraulically powered; hydraulic pressure is produced from a pump driven off the power unit. There is a manual back-up pump fitted. The front suspension is hydraulically locked for firing. One man can, in emergency, deploy and operate the gun! It is possible to fire the gun in the travelling configuration.

The engine is a 1500 cc Porsche which gives a top speed of 20 km/hr; gradients of 50 per cent can be climbed. Manoeuvrability and cross country performance are good: the main drive wheels can be brake steered, and the rear wheels are also steerable.

Employment Germany (FRG), trials only

The gun in action

With barrel retracted and ready to move off under its own power

90 mm recoilless rifle PV 1110

Sweden

Calibre	90 mm
Barrel length	36.9 cm
Muzzle brake	Nil
Trail type	Pole
Weight	260 kg
Ammunition	HEAT (3·1 kg)
Armour penetration	380 mm
Rate of fire	6 rds/min
Muzzle velocity	715 m/sec
Maximum range	—
Maximum effective range	900 m
Elevation limits	−10° + 15°
Traverse on carriage	75°
Detachment	2–3

The PV 1110, made by Bofors, is a smooth bore weapon firing a fin stabilised hollow charge projectile of great penetrating power. The weight of a complete round is 9·6 kilogrammes. The sustained rate of fire is six rounds/minute, and two rounds can be fired in 13 seconds. A 7·62 mm spotting rifle with magazine of 10 rounds is mounted over the barrel. A pistol grip and trigger is fitted below the barrel. The overall weight varies with the mounting, but the gun itself weighs 125 kg. Traverse also varies with the mounting.

The PV 1110 has three different modes of transport. The normal two-wheel trailer carriage can be replaced by a simple sledge, or the weapon can be vehicle mounted. The trailer consists of a broad sheet steel platform with a two wheel axle and single steady leg by which the weapon is towed. The sledge is a long slender platform on which the piece is mounted so low that the detachment must lie down to operate the weapon: the overall height is only 470 mm. Not only does this give it the lowest profile of any anti-tank weapon of its type, but it obviates the difficulties of digging-in in snowbound country. The SP is mounted on the Volvo 4 × 4 Laplander. The same vehicle, or the Austrian Haflinger, are used as tractors for the towed version.

In each of these the barrel is mounted at its centre of gravity in a tubular member shaped like a question mark.

Employment Eire Sweden

This action photograph demonstrates the extremely low silhouette of the weapon

90 mm anti-tank gun Model 50

Switzerland

Calibre	90 mm
Barrel length	—
Muzzle brake	None
Trail type	Split
Weight	631 kg
Ammunition	HEAT (1·95 kg)
Armour penetration	250 mm
Rate of fire	12–15 rds/min
Muzzle velocity	600 m/sec
Maximum range	—
Maximum effective range	650 m
Elevation limits	—
Traverse on carriage	—
Detachment	5–6

There are two types of 90 mm Swiss anti-tank gun which are very similar. Both have the same downward curved and fragile looking box section trail leg design.

The recoil system is mounted below the barrel. The breech block is an upward opening vertical sliding wedge. No performance figures are available for the ammunition, but its muzzle velocity suggests that it has something in common with the Model 57.

The trail legs of the two weapons end in different types of spade. A simple plank seat is provided for the layer on the left trail leg.

The chief distinguishing feature is the shield. That of the Model 50 is square and is flat in its lateral axis: it bends back at the top and forward to clear the wheels at the bottom.

Employment Switzerland

90 mm anti-tank gun Model 57

Switzerland

Calibre	90 mm
Barrel length	—
Muzzle brake	Multi-baffle
Trail type	Split
Weight	716 kg
Ammunition	HEAT (2·7 kg)
Armour penetration	250 mm
Rate of fire	12–15 rds/min
Muzzle velocity	600 m/sec
Maximum range	—
Maximum effective range	800–1,000 m
Elevation limits	—
Traverse on carriage	—
Detachment	5–6

This picture shows the high mounting of the sights, the breech protecting bar and the hooded shield. Note also that the wheels hinge with the trail legs when they are opened in action

The Model 57 is an improved version of the earlier Model 50. It has the same elegant downward curve to the trail legs, but is easily distinguished by its new shield. It is also known as the Pak 57.

The long barrel is mounted at the centre of gravity, well forward of the breech. The vertically sliding breech block opens downwards. A bar arrangement beneath the breech protects it from damage against ground on recoil at high angles of elevation. The forward mounting of the trunnions and consequently large vertical movements of the breech makes this necessary. The recoil system is mounted under and over the barrel. Access to the breech is good, allowing a high rate of fire.

The sights are mounted in an upward extension of the saddle. The wooden bench seat of the Type 50 is replaced by a shaped seat for the layer.

The wheel axles are fitted to the trail legs and hinge with them on deployment. The shield is well rounded at the top and sweeps back from the front of the recoil system in the form of a hood over the sights.

Employment Switzerland

Recoilless rifle M 58

Finland

Calibre	95 mm
Barrel length	34·5 calibres
Muzzle brake	Nil
Trail type	Pole
Weight	140 kg
Ammunition	HEAT (10·1 kg)
Armour penetration	300 mm
Rate of fire	6–8 rds/min
Muzzle velocity	615 m/sec
Maximum range	1,000 m
Maximum effective range	700 m
Elevation limits	—
Traverse on carriage	—
Detachment	3

95 mm recoilless rifle with crew at their posts

This is the only recoilless gun of 95 mm calibre. It is a purely Finnish design. It has two small wheels, but lifting handles are fitted front and rear (forward of the breech) to assist in manhandling. Its detachment of only three can carry it over short distances.

There are two stabilizing legs fitted front and rear.

Used in the heavy weapons companies of infantry battalions.

Employment Finland

---

100 mm field anti-tank gun M 1944 (D 10)

Union of Soviet Socialist Republics

Calibre	100 mm
Barrel length	54 calibres
Muzzle brake	Double-baffle
Trail type	Split
Weight	3,460 kg
Ammunition	HVAP (9·38 kg)
	HEAT (12·2 kg)
	APHE (15·9 kg)
	HE (15·7 kg)
Armour penetration	380 mm (HEAT)
	153 mm (APHE)
Rate of fire	8–10 rds/min
Muzzle velocity	1,000 m/sec (APHE)
	900 m/sec (HE)
	800 m/sec (HEAT)
	1,100 m/sec (HVAP)
Maximum range	21,000 m
Maximum effective range	1,000 m
Elevation limits	$-5° +40°$
Traverse on carriage	55°
Detachment	6

100 mm M 1944 in the field role

The barrel of this equipment was designed for the Russian M 1939 Anti-Aircraft Gun. It appeared as the D 10 at the end of World War II and has since been developed into the M 1955. A D 10T version was mounted in the T 54 tank and SU 100 assault gun. This weapon then is the middle one of three generations of development.

It is essentially an anti-armour weapon despite its name: its ammunition options, low silhouette, limited elevation and high muzzle velocity all indicate this. Fixed ammunition and semi-automatic vertically sliding breech permit a high rate of fire. The ammunition has an excellent armour penetrating performance. The provision of HE ammunition gives the weapon its dual role. Double wheels are fitted to a simple split trail of box section legs. The shield is well angled and mounts useful storage bins under the barrel.

The recoil system is mounted below the barrel.

As is common practice these weapons were passed on to Soviet allies when the M 1955 was introduced. M 1955's have in turn been exported and it is not always clear which country has this gun and which the M 1955.

Employment			
	Bulgaria	Mongolia	Sudan
	China	Pakistan	USSR
	Congo	Poland	Vietnam
	Egypt	Rumania	Yugoslavia
	India	Somalia	
	North Korea	Syria	

100 mm field anti-tank gun M 1944 (D 10)

100 mm field anti-tank gun M 1955

Union of Soviet Socialist Republics

Calibre	100 mm
Barrel length	54 calibres
Muzzle brake	Pepper-pot
Trail type	Split
Weight	2,700 kg
Ammunition	HVAP (9·38 kg)
	HEAT (10·2 kg)
	APHE (15·9 kg)
	HE (15·7 kg)
Armour penetration	380 mm (HEAT)
	153 mm at 500 m (APHE)
Rate of fire	10 rds/min
Muzzle velocity	1,000 m/sec (APHE)
	900 m/sec (HE)
	800 m/sec (HEAT)
	1,100 m/sec (HVAP)
Maximum range	21,000 m
Maximum effective range	1,000 m
Elevation limits	−5° + 40°
Traverse on carriage	55°
Detachment	8

This long barrelled gun is variously described as an anti-tank and field gun. It is used in both roles, has been mounted in T 54A and T 55 tanks and given a naval mounting. It was especially successful during the Yom Kippur War as an anti-tank gun.

It replaced the M 1944 (D 10) which was similarly a multi-role equipment. The ammunition system is that of the older gun, and shell options centre on an anti-armour capability. The HEAT round is especially effective. Ammunition is fixed and there is no provision for clearing difficult crests either by charge reductions or high angle fire.

The APN 3–5 infra-red sight is fitted; and a new

image intensifying sight is available.

The M 1955 can be distinguished from the earlier D 10 by its single road wheels, castor wheel on the trail and pepper-pot muzzle brake.

The recoil system is fitted above the barrel and behind the shield. It is 600 kg lighter than the D 10. A modified version with elevation limits of −10° and +20° is known as the T 12.

Employment

Bulgaria	India	Somalia
China	Iraq	Syria
Czechoslovakia	North Korea	USSR
Egypt	Mongolia	Vietnam
Germany (GDR)	Poland	Yugoslavia
Hungary	Rumania	

100 mm M 1955 field anti-tank gun

---

105 mm recoilless rifle M 65

Yugoslavia

Calibre	105 mm
Barrel length	40 calibres
Muzzle brake	Nil
Trail type	Pole
Weight	280 kg
Ammunition	HEAT
Armour penetration	330 mm
Rate of fire	6 rds/min
Muzzle velocity	—
Maximum range	6,000 m
Maximum effective range	600 m
Elevation limits	—
Traverse on carriage	—
Detachment	—

Whereas the Soviet Army has now given up large-calibre recoilless guns, Yugoslavia, with its problem of finding effective weapons for use in mountainous terrain, has developed its own large recoilles rifle, in addition to two lighter anti-tank weapons, the RB 57 and the M 60.

Among communist countries' weapons, the M 65 is unusual in that it is fitted with a 12·7 mm spotting machine gun, the UB, used at ranges up to 600 m (the M 65's maximum effective anti-tank range). It employs a 20 round belt, fired either automatic or single shot. The M 65 is a breech-loading weapon with a very long rifled barrel, allowing the gun to fire its HEAT projectiles a distance of 6,000 m. HE rounds have not yet been seen with this weapon. Six rounds can be fired in one minute and the shaped-charge warhead will pierce 330 mm of armour plate.

The gun is mounted on a tripod, the two wheels attaching to the base of the two legs, whilst the third leg extends to the rear and also acts as a towbar. A small, square shield which protects the firer's face from blast is fitted in front of the optical sight. This weapon can travel in the normal manner towed by a truck or armoured personnel carrier, or can be broken down for carriage by mule.

105 mm recoilless rifle M 65

Employment Yugoslavia

106 mm recoilless rifle M 40 A1

United States of America

Calibre	106 mm
Barrel length	26 calibres
Muzzle brake	Nil
Trail type	—
Weight	130 kg
Ammunition	HEAT (7·71 kg) HEP-T (7·71 kg)
Armour penetration	450 mm (HEAT)
Rate of fire	5 rds/min
Muzzle velocity	503 m/sec (HEAT) 498 m/sec (HEP-T)
Maximum range	7,700 m
Maximum effective range	1,000 m
Elevation limits	−17° + 65°
Traverse on carriage	360°
Detachment	2

This popular weapon dates from 1953. It is a conventional rifled, breech loaded weapon, with an interrupted screw type breech. Barrel life is in excess of 2,500 rounds. Percussion firing is used. A ·50 calibre (12·7 mm) spotting rifle is mounted above the barrel. The basic weapon less mounting weighs only 130 kilogrammes.

Elevation varies with the mounting, but is very generous for the trailer mount. At 39·5° elevation the HEP-T round ranges to 6,876 m, giving the weapon a useful secondary role if the necessary fire control arrangements can be made. Armour penetration is formidable. A modified weapon, the M 40AIC has an improved breech mechanism.

A number of mounts have been developed. These are:

M 27: A plain short-legged tripod with a 360° traversing ring.

M 79: Three flat, box-girder legs attached horizontally to a central pedestal with a castor wheel on the forward leg. A large elevating wheel is fitted to the top of the pedestal. This mount is used to fix the weapon into a ¼ ton Jeep, which is the most popular configuration.

M 92: A pillar type mount designed to fit the weapon to an open platform mechanical mule for the U.S. Marine Corps.

M 149-E5: A multiple mounting for six rifles fitted to a tracked carrier (ONTOS), which is no longer in service with US forces, but is still in use by Guatamala, Lebanon, Morocco and Venezuela.

The M 40AI is built under licence in Brazil, Israel, Japan (known as Type 60) and Spain. The Austrian M 40AI's are known as Pak 66 and are fitted to a locally designed trailer—giving two barrel positions which give an overall height of 0·63 m or 0·93 m.

Employment

Australia	Greece	South Korea	Portugal
Austria	India	Lebanon	Singapore
Brazil	Indonesia	Liberia	Spain
Cameroun	Iran	Luxemburg	Switzerland
Canada	Israel	Netherlands	Taiwan
Chile	Italy	New Zealand	Thailand
Denmark	Japan	Norway	Turkey
France	Jordan	Pakistan	USA
Germany (FRG)	Kampuchea	Philippines	Vietnam

The 106 mm recoilless rifle M 40 A1 mounted on a jeep

107 mm recoilless rifle B-11

Union of Soviet Socialist Republics

Calibre	107 mm
Barrel length	33 calibres
Muzzle brake	Nil
Trail type	Tripod
Weight	305 kg
Ammunition	HEAT (9 kg) HE (13·6 kg)
Armour penetration	380 mm at 450m (HEAT)
Rate of fire	5–6 rds/min
Muzzle velocity	410 m/sec
Maximum range	6,650 m
Maximum effective range	450 m
Elevation	−10° + 45°
Traverse on carriage	35°
Detachment	5

The B-11 consists of the barrel, a simple tripod and a two wheel axle. For firing the wheels can be removed. It weighs only 240 kg.

It is a smooth bore equipment, firing spin stabilised HE and HEAT ammunition. The heavier HE round can be fired in an indirect fire role to 6,650 m, but in its primary anti-tank role the maximum effective range is 450 m. Armour penetration of 360 mm is at a 90° impact angle. The PBO 4 sights are mounted in a gimbal above the barrel, and are used for direct and indirect laying.

The breech hinges down for loading. The chamber is jacketed with a grill to protect the detachment from the heat generated. It is towed by the barrel.

Traverse is limited, but the weight of gun and size of detachment make manhandling easy.

It is no longer in service in the Soviet Army.

Employment

Bulgaria	Germany (GDR)	North Korea
China	Hungary	Laos
Egypt	Kampuchea	Vietnam

B-11s firing

76 mm SP anti-tank gun SU 76

Union of Soviet Socialist Republics

Calibre	76·2 mm
Barrel length	42·6 calibres
Muzzle brake	Double baffle
Ammunition	HVAP (3·1 kg)
	APHE (6·5 kg)
	HEAT (5·3 kg)
	HE (6·2 kg)
Armour penetration	92 mm at 500 m (HVAP)
	69 mm at 500 m (APHE)
	120 mm (HEAT)
Rate of fire	20 rds/min
Muzzle velocity	965 m/sec (HVAP)
	680 m/sec (HE)
	746 m/sec (APHE)
	325 m/sec (HEAT)
Maximum range	12,200 m
Maximum effective range	500 m
Elevation limits	−5° + 34°
Traverse on mounting	30°
Detachment	4
Chassis—Type	T 70
Engine	2 × Otto petrol
Power	2 × 85 bhp
Speed	45 km/hour
Endurance	230 km
Ammunition carried	60
Height × length × width	2·3 × 5·1 × 2·73 m
Weight	11,200 kg

This is the ZIS 3 gun mounted on a lengthened T 70 tank chassis and introduced in 1943. Despite its age it is used extensively outside the Soviet Union. The anti-armour performance is poor by modern standards, but remains effective against APCs.

It is unlikely that the HE round is much used in view of the equipment's primary role as a front-line tank destroyer. However its availability adds flexibility to its deployment. 60 rounds are carried on board—a very good supply.

The PT 70 chassis is powered by two petrol engines each delivering 85 bhp. These give a reasonable 45 km/hr road speed, but the vehicle has a somewhat limited range of action. It is fitted with very narrow tracks running on six road wheels, but the low weight and length of track keep ground pressure down to 0·6 kg/cm². The driver sits in the centre.

Employment Albania
China
North Korea
Vietnam

76 mm SPs—crews resting

85 mm SP anti-tank gun ASU 85

Union of Soviet Socialist Republics

Calibre	85 mm
Barrel length	55 calibres
Muzzle brake	Double baffle
Ammunition	HVAP (5 kg) APHE (9·3 kg) HE (9·5 kg)
Armour penetration	113 mm at 1,000 m (HVAP) 91 mm at 1,000 m (APHE)
Rate of fire	4 rds/min
Muzzle velocity	1,030 m/sec (HVAP) 792 m/sec (HE and APHE)
Maximum range	—
Maximum effective range	1,000 m
Elevation limits	−4° +15°
Traverse on mounting	12°
Detachment	4
Chassis—Type	PT 76
Engine	V-6 diesel
Power	240 bhp
Speed	45 km/hour
Endurance	260 km
Ammunition carried	40
Height × length × width	2·13 × 6 × 2·8 m
Weight	14,000 kg

The ASU 85 dates from 1962, but was first shown in 1964 in Poland. It consists of the D 44 gun mounted on the PT-76 amphibious tank chassis.

The gun itself entered service during World War II, but is still widely used in a number of roles including anti-tank and anti-aircraft. The long barrel is fitted with a vertically sliding breech. Armour penetration is poor by modern standards, but more than adequate to defeat armoured personnel carriers or the side armour of tanks.

Like the ASU 57 the ASU 85 is designed primarily for airborne or air-transported operations, although it is too heavy to be air-dropped or carried by helicopter.

The rate of fire is considerably less than that of the towed version because the detachment is smaller and the enclosed fighting compartment constricts movement.

The PT 76 is a popular chassis for hybrid weapons. It is light but roomy and has inherent buoyancy. The V-6 engine gives a reasonable road speed, and hydro-jets propel the chassis in water at 10 km/hour.

Employment Germany (GDR), Poland, USSR

ASU 85s of an airborne formation marching past

90 mm SP anti-tank gun JPz 4-5

Federal Republic of Germany

Calibre	90 mm
Barrel length	40·4 calibres
Muzzle brake	Double baffle
Ammunition	HEAT HEP-T HVAP Practice
Armour penetration	330 mm (HEAT)
Rate of fire	12 rds/min
Muzzle velocity	1,181 m/sec (HEAT)
Maximum range	15,000 m
Maximum effective range	2,000 m
Elevation limits	−8° + 15°
Traverse on mountings	30°
Detachment	4
Chassis—Type	Individual
Engine	Daimler Benz 837 Aa diesel
Power	500 bhp
Speed	70 km/hour
Endurance	400 km
Ammunition carried	51
Height × length × width	2·08 × 6·23 (less gun) × 2·98 m
Weight	26,000 kg

Production of 800 of these anti-tank weapons finished in 1967. The 90 mm Bord K L/40·8 gun carries a double baffle muzzle brake and fume extractor. Few details of the gun's performance have been published.

An infra-red target search light, and associated sights, is available, and the driver can be provided with an infra-red viewer. Sighting and laying gear are conventional and range is determined by commander's estimate.

The gun is mounted on the front glacis plate of the hull and is provided with muzzle brake and fume extractor. There is inboard stowage for 51 rounds.

The chassis runs on five road wheels with three track return rollers. It is powered by a Daimler Benz 837 Aa engine. This is a V8 four stroke diesel producing 500 bhp and, with a power to weight ratio of 19·5 bhp/ton, gives remarkably agile performance. The transmission is the Renk HSWL 123, three speed automatic, fitted with forward and reverse transfer box: performance is therefore the same in both directions.

The weapon is the standard equipment of Panzer-Grenadier Battalions.

Employment Belgium
Germany (FRG)

90 mm SP anti-tank gun JPz 4-5

90 mm Tank Destroyer IKV 91

Sweden

Calibre	90 mm
Barrel length	54 calibres
Muzzle brake	Nil
Ammunition	HEAT (4·5 kg) HE (6·7 kg)
Armour penetration	—
Rate of fire	—
Muzzle velocity	845 m/sec (HEAT) 600 m/sec (HE)
Maximum range	—
Maximum effective range	1,500 m
Elevation limits	−10° + 15°
Traverse on mounting	360°
Crew	4
Chassis—Type	Full track
Engine	Turbocharged diesel
Power	330 bhp
Speed	70 km/hr
Endurance	550 km
Ammunition carried	59
Height × length × width	2·35 × 6·41 × 3·0 m
Weight	16,000 kg

The IKV 91 can function equally well as a tank destroyer or as a light tank, and its specified function is the support of infantry. It entered Swedish service late in 1975.

The gun, specially designed for this application by AB Bofors and designated by them KV 90 S73, is a low-pressure type which thus produces low recoil stresses and minimum muzzle blast and target obscuration. The gun is smoothbored and fires fin-stabilised projectiles. It is fitted with a fume extractor. Little information regarding its performance has been released.

The vehicle is full-tracked with torsion bar supension, a turbocharged Volvo-Penta diesel engine, and an automatic Allison transmission. Turret traverse and gun elevation are hydraulically powered. Fire control is by electronic computer and laser rangefinder; the rangefinder instructs the computer, which then automatically drives the elevation and traverse motors to bring the gun to bear on the target, the correct amounts of lead, aim-off and other corrections having been applied in the computer. Tests have shown that with the turret 30° away from a target, the entire sequence of acquiring, computing, laying and firing can be done in nine seconds.

Employment Sweden

The IKV 91 tank destroyer. It is fully amphibious, and the folded trim vane can be seen at the hull front

100 mm SP anti-tank gun SU 100

Union of Soviet Socialist Republics

Calibre	100 mm
Barrel elngth	54 calibres
Muzzle brake	Nil
Ammunition	HVAP HEAT APHE (15·9 kg) HE (15·7 kg)
Armour penetration	380 mm (HEAT) 153 mm at 500 m (APHE) 180 mm at 500 m (HVAP)
Rate of fire	6–8 rds/min
Muzzle velocity	1,000 m/sec (APHE) 900 m/sec (HE and HEAT) 1,100 m/sec (HVAP)
Maximum range	15,400 m
Maximum effective range	1,000 m
Elevation limits	$-2° + 17°$
Traverse on mounting	16°
Detachment	4
Chassis—Type	T 34
Engine	12 cylinder water cooled diesel
Power	530 bhp
Speed	55 km/hour
Endurance	300 km
Ammunition stowage	34
Height × length × width	2·24 × 9·45 × 3·05 m
Weight	31,600 kg

This equipment consists of the D 10 gun mounted on the T 34/85 tank chasis.

The SU (Samokhodnaya Ustanovka) series of Soviet SP anti-tank guns are designed to fit any given tank chassis with a heavier gun than is possible in a turret. The penalty is a severe restriction in traverse, so that the tactical handling of these weapons is complementary with its corresponding tank; the SU85 with T-34/76 and this gun with T-34/85.

Only minor modifications have been made to the gun. The detachment is reduced to four, which must have an adverse effect on the rate of fire (8–10 rounds/minute) which is quoted for the towed gun. An HE round is available for the towed gun which has a recognised field role. While the SU 100 will accept this ammunition, it is unlikely that it will be carried in the on-board first line ammunition if the weapon is to be exploited fully as a tank destroyer.

The gun is mounted in a convex mantle to the right of the driver. A 'conning tower' is provided for the commander on the right of the casemate.

The T 34/85 chassis carries 75 mm of frontal armour. Its diesel engine develops 530 bhp and gives a good top speed of 55 km/hr. Ground pressure is 0·8 kg/cm^2.

Employment

Albania	Egypt	Morocco
Algeria	Germany (GDR)	Rumania
Bulgaria	Iraq	Syria
China	North Korea	North Yemen
Cuba	Mongolia	Yugoslavia
Czechoslovakia		

100 mm SP anti-tank gun SU-100

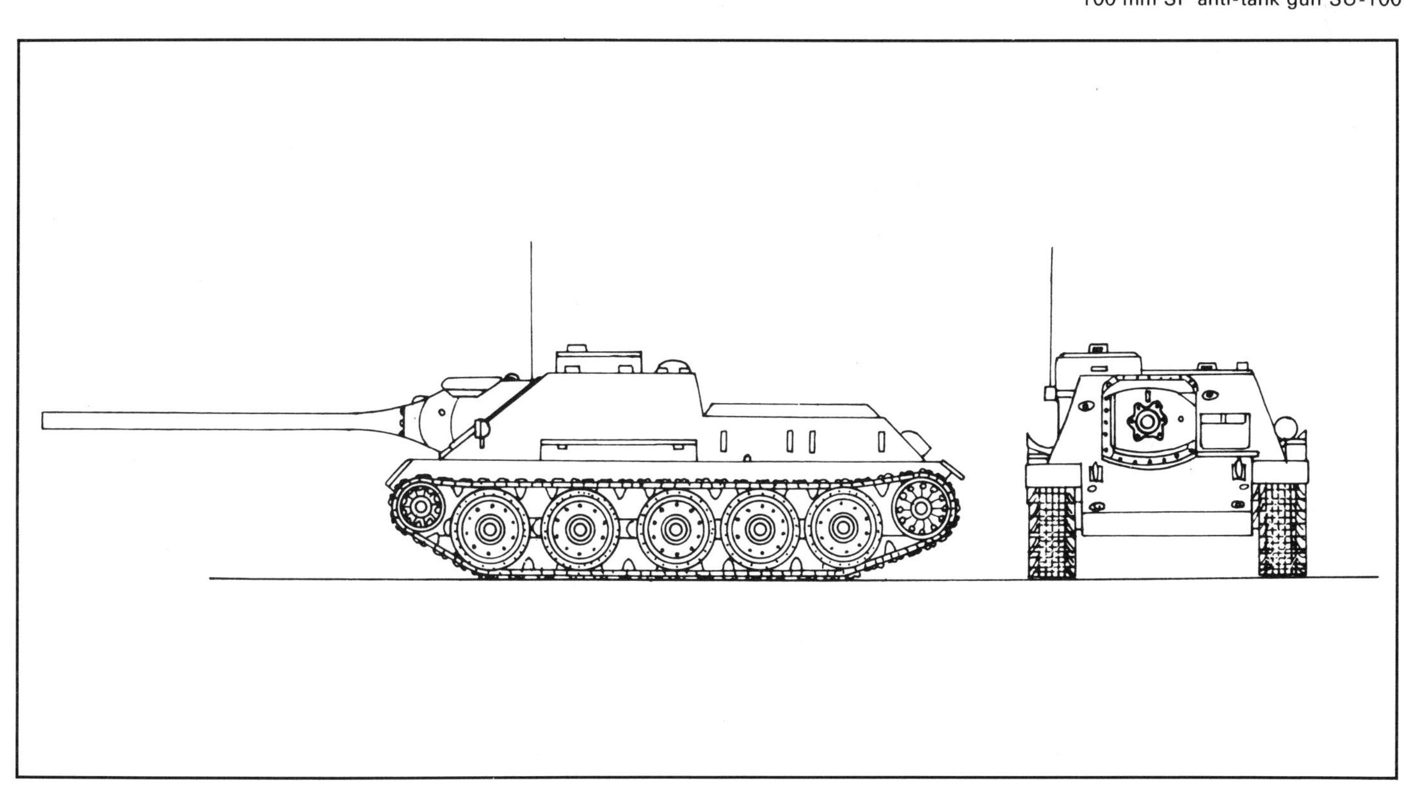

105 mm SP anti-tank gun 'K'

Austria

Calibre	105 mm
Barrel length	44 calibres
Muzzle brake	Single baffle
Ammunition	HEAT APHE Smoke
Armour penetration	360 mm (HEAT)
Rate of fire	12 rds/min
Muzzle velocity	800 m/sec (HEAT) 700 m/sec (APHE)
Maximum range	2,000 m
Maximum effective range	—
Elevation limits	−6° + 13°
Traverse on mounting	360°
Detachment	3
Chassis—Type	Saurer 4K
Engine	Saurer 4FA
Power	300 bhp
Speed	63 km/hour
Endurance	—
Ammunition carried	17
Height × length × width	1·97 × 5·57 (less gun) × 2·5 m
Weight	17,000 kg

The gun is the French 105 mm D 1504, which is fitted to AMX-13 tank. It fires HEAT and APHE ammunition. Loading is semi-automatic from two six-round magazines in the turret. There is provision for a laser range finder. The X7·5 sight is fitted, together with a dual magnification sight for the commander.

The outstanding feature of this gun is the French FL 12 'oscillating' turret. It consists of a traversing ring which has a pent-roof side elevation, at the apex of which the elevating mass is mounted in trunnions. The result is that the gun is mounted almost level with the top of the turret. A rearward turret extension accommodates recoil and the two magazines. The recoil system is entirely under armour. The turret is power traversed, and will rotate through 360° in 12 seconds.

The chassis is from the Saurer 4K APC, modified to drive the rear sprocket. There are five road wheels and three return rollers. Hydraulic dampers are fitted to the front and rear wheels only. The engine is an in-line six cylinder diesel producing 300 bhp. A manual five-speed gearbox (2F 655) is used. The vehicle wades to one metre unprepared.

Employment Austria

---

106 mm SP recoilless rifle Type 60

Japan

Calibre	106 mm
Barrel length	3,400 mm
Muzzle brake	Nil
Ammunition	HEAT (7·71 kg) HEP-T (7·71 kg)
Armour penetration	450 mm (HEAT)
Rate of fire	6 rds/min
Muzzle velocity	503 m/sec (HEAT) 498 m/sec (HEP-T)
Maximum range	7,700 m
Maximum effective range	1,000 m
Elevation limits	−5° + 10° (see text)
Traverse on mounting	20° (see text)
Detachment	3
Chassis—Type	Individual
Engine	Komatsu 6-cylinder diesel
Power	120 bhp
Speed	48 km/hr
Endurance	130 km
Ammunition carried	10
Height × length × width	1·38 × 4·3 × 2·23 m
Weight	8,020 kg

This weapon entered service in 1960 as the first Japanese post-war AFV. It consists of two recoilless anti-tank rifles mounted on a specially designed tracked chassis. The rifles are the American M 40 A1 weapon made under licence in Japan. The vehicle is similar in concept to the obsolete ONTOS, a US Marine Corps weapon which mounted six rifles.

The smooth bore, breech loaded weapon has an interrupted screw type breech and uses percussion firing. Armour penetration is formidable.

A single ·50 calibre (12·7 mm) spotting machine gun is used to augment the optical sights. No night fighting aids are fitted. A two position mounting gives either an exceptionally low silhouette or, with the guns raised, an increased arc of fire: traverse increases to a total of 60°, and elevation limits are −20° and +15°.

The mount does not have to accept the shock of recoil associated with conventional guns, and a light chassis can be employed. The small six cylinder air cooled diesel engine produces a good turn of speed, while ground pressure is kept down to 0·63 kg/cm². The vehicle wades to 0·8 m.

Employment Japan

152 mm assault gun JSU 152

Union of Soviet Socialist Republics

Calibre	152 mm
Barrel length	32 calibres
Muzzle brake	Multi-baffle
Ammunition	APHE (48·8 kg) HE (43·6 kg)
Armour penetration	124 mm at 500 m (APHE)
Rate of fire	—
Muzzle velocity	655 m/sec (HE) 600 m/sec (APHE)
Maximum range	9,000 m
Elevation limits	−3° + 20°
Traverse on mounting	20°
Detachment	5
Chassis—Type	JS 2
Engine	Diesel
Power	520 bhp
Speed	37 km/hour
Endurance	150 km
Ammunition carried	20
Height × length × width	2·45 × 9·05 × 3·04 m
Weight	46,500 kg

During World War II the Russians produced a 152 mm SP assault gun on the KV tank chassis: this was the SU 152, and was superseded after the war by the JSU 152.

This gun is a derivative of the M 1937 (ML 20) howitzer mounted on the JS 2 chassis. It is employed as a heavy anti-tank weapon, assault gun, and SP howitzer, and is thus a very versatile weapon.

The gun projects from a mantlet which is shaped to cover the recoil mechanism mounted below the barrel. There is no turret, the super-structure being built up into a casemate. Only 20 rounds of ammunition are carried.

The chassis runs on six road wheels. Its endurance is very limited, but auxiliary fuel tanks can be mounted either side of the rear mounted engine's decking.

It is the heaviest and largest SP gun to be produced in Russia. Its smaller relative the JSU 122 is no longer in service, but this 152 mm weapon is still used in a few countries.

Employment Algeria
Czechoslovakia
Egypt
Iraq

Assault gun JSU 152

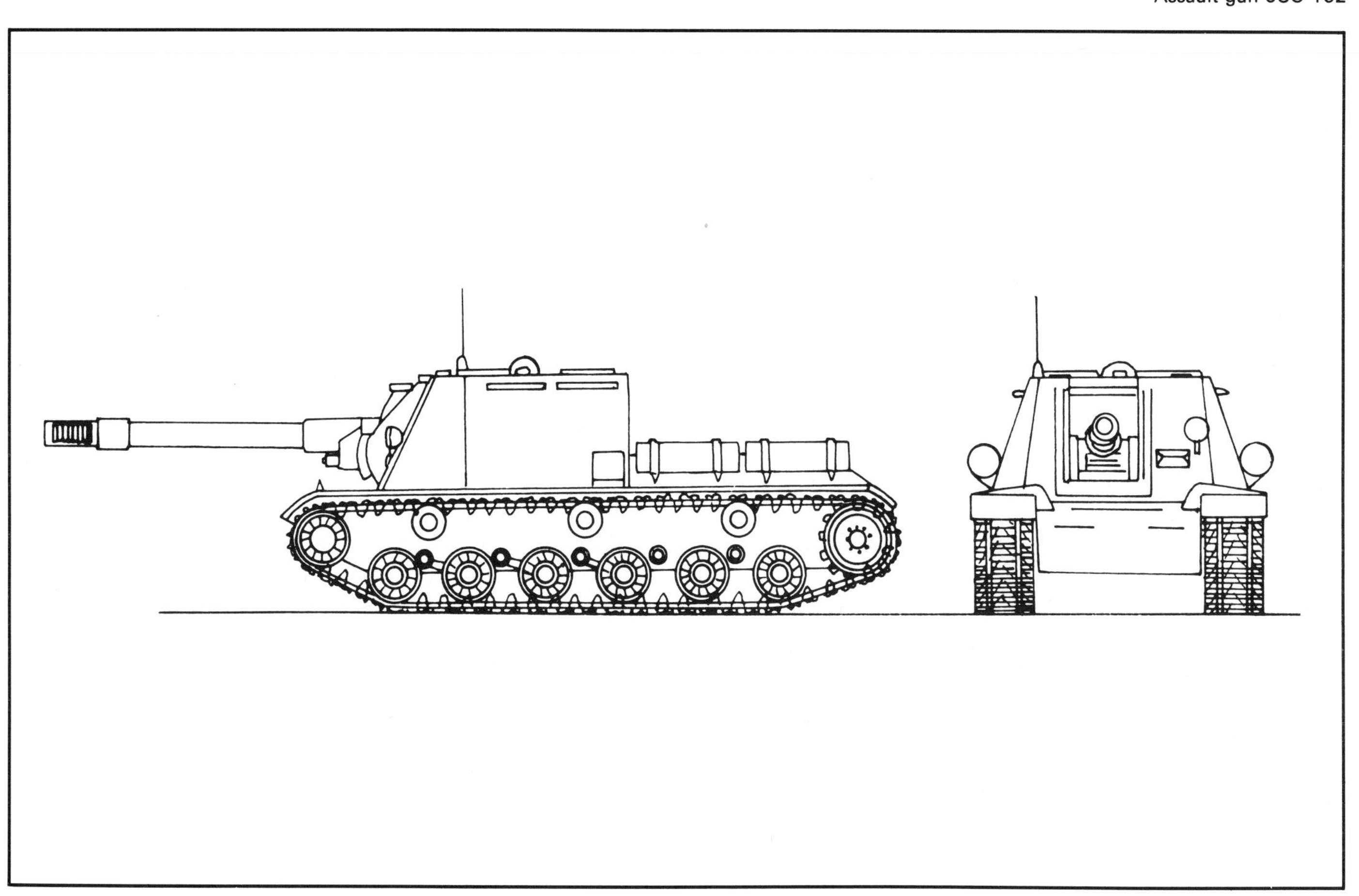

Union of Soviet Socialist Republics

Missile diameter	120 mm
Missile length	880 mm
Missile weight (before launch)	11·3 kg
Missile wing span	460 mm
Guidance system	Manual command to line of sight
Propulsion	Two stage solid propellant
Velocity	120 m/sec
Range—Maximum	3,000 m
Minimum	500 m
Time of flight to maximum range	25 secs
Warhead	Shaped charge

Sagger on BRDM, showing launcher position

Sagger has now replaced Snapper in Soviet equipped Armies. It can be mounted on BMP or BRDM, or it can be manpacked and fired from the ground. In the BMP APC version the missile rail is mounted above the 73 mm smooth bore gun and can be fired from within the vehicle. On the BRDM reconnaissance vehicle six Sagger missiles are mounted on a row of launcher rails which can be elevated to the firing position from inside the vehicle. For manpacked use it has a team of three who can set out up to four missiles connected to a single control sight, with the lid of the carrying case for each missile acting as a launching stand. In the Middle East War of 1973 Sagger achieved fame in the world press as the 'suitcase missile'. The HEAT warhead has a penetration of better than 400 mm of armour plate.

Employment

Afghanistan	Germany (GDR)	Rumania
Algeria	Hungary	Syria
Angola	Iraq	Uganda
Bulgaria	Israel	USSR
Czechoslovakia	Libya	Vietnam
Egypt	Mozambique	Yugoslavia
Ethiopia	Poland	

Sagger mounted on BMP tracked reconnaissance vehicle

Sagger

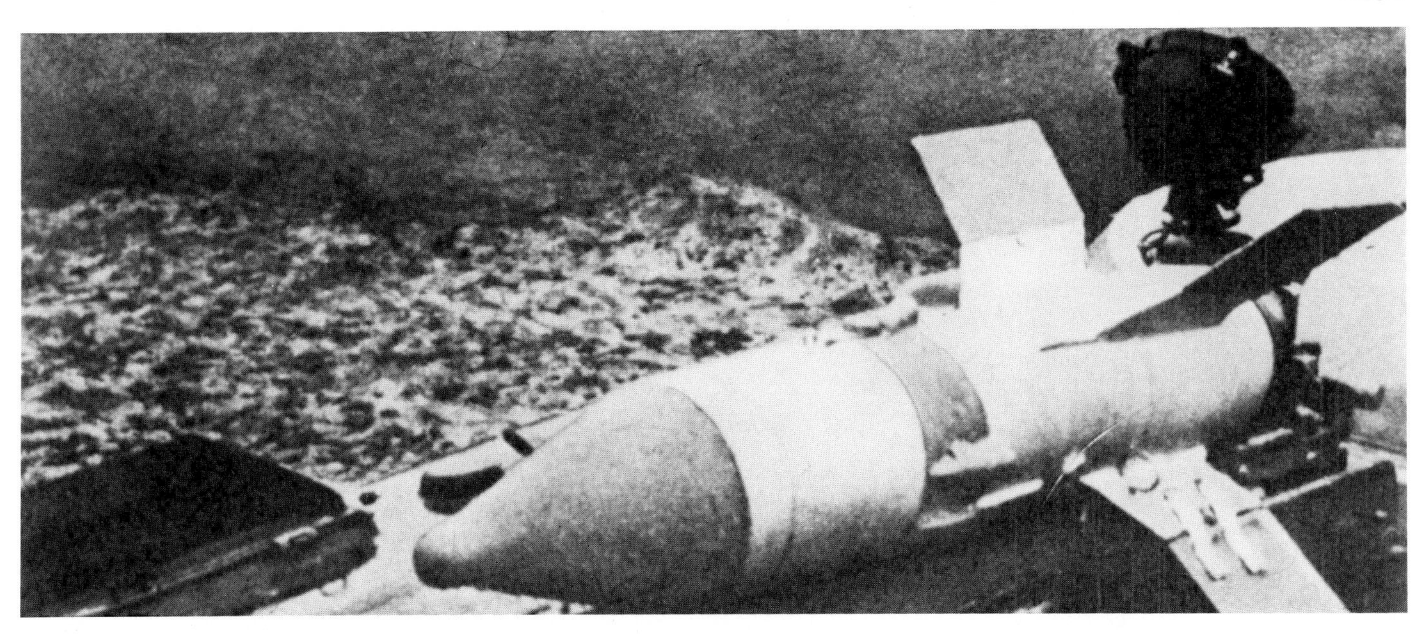

AT-4 (NATO code—Spigot)

Union of Soviet Socialist Republics

This replacement for the AT-3 Sagger is fired from a tube launcher mounted on a tripod in a similar fashion to the US TOW. The missile is operational with Soviet forces, but no further details are known.

Swatter anti-tank guided weapon

Union of Soviet Socialist Republics

Missile diameter	150 mm
Missile length	900 mm
Missile weight (before launch)	25 kg
Missile wing span	650 mm
Guidance system	Manual command to line of sight
Propulsion	Two-stage solid propellant
Velocity	150 m/sec
Range—Maximum	2,200 m
Minimum	300 m
Time of flight to maximum range	15 sec
Warhead	Shaped charge

Swatter is normally mounted in the BRDM reconnaissance behicle in a row of three launchers, which can be elevated from its normal travelling position inside the vehicle to the firing position. It is a manual system, but it is the only anti-tank guided weapon to use a radio command link between the tracker sight and the missile. Although this avoids the problem of wire breaks which still occasionally occur in the wire guided system, it does make Swatter vulnerable to electronic counter measures. The HEAT warhead has a delayed action fuze, and it is said to achieve about 500 mm penetration of armour plate.

(illustrated overleaf)

Employment		
	Czechoslovakia	Poland
	Egypt	Rumania
	Germany (GDR)	Syria
	Hungary	USSR

Swatter, mounted on BRDM 1 reconnaissance vehicle

AT-5 (NATO code—Spandrel)

Union of Soviet Socialist Republics

AT-5 is fired from launch tubes mounted on BRDM armoured cars, each of which has five launchers fitted. This combination of missile and vehicle is expected to replace mobile versions of the earlier AT-2 Swatter and AT-3 Sagger. AT-5 uses semi-automatic command to line-of-sight guidance, commands being transmitted down trailing wires. Little else is known about it.

AT-6 (NATO code—Spiral)

Union of Soviet Socialist Republics

AT-6 has been developed to arm the Mil Mi-28 Hind D attack helicopter, in which role it replaces the AT-2 Swatters used as interim weapons. Semi-active laser guidance is reported to be used, allowing the missile to home on to laser energy transmitted by ground vehicles and reflected from targets. A range of up to 10 km may be possible.

SS 11 anti-tank guided weapon

France

Missile length	1·21 m
Missile diameter	164 mm
Missile weight (before launch)	29·9 kg
Missile wing span	500 mm
Guidance	Manual command to line of sight
Propulsion	Two stage solid propellant
Velocity	150 m/sec
Range—Maximum	3,000 m
Minimum	300 m
Time of flight to maximum range	20 sec
Warhead	Shaped charge/HEAT/ fragmentation

SS 11 has twice the range of the earlier SS 10. It is normally fired from a vehicle, but also from a ground launcher. It is manually controlled by a thumbstick, which transmits guidance commands through a wire link to the missile. The missile is automatically gathered into the field of view of the optical sight and is then flown on to the target by the operator. Its HEAT warhead can penetrate 600 mm of armour. The SS 11 is one of the best of the first generation anti-tank guided weapons and it is still widely used in Western Armies, as well as in the Middle East and India. It is gradually being replaced by SS 12, HOT or TOW.

Employment

Argentina	Iraq	Saudi Arabia
Belgium	Israel	Spain
Canada	Italy	Sweden
Chile	Kuwait	Switzerland
Denmark	Lebanon	Tunisia
Germany (FRG)	Libya	Turkey
Finland	Malaysia	Uganda
France	Norway	United Arab Emirates
Greece	Pakistan	Venezuela
India	Peru	
Iran	Portugal	

SS 11 missile mounted on West German RJPz armoured personnel carrier

SS 12 anti-tank guided weapon

France

Missile diameter	210 mm
Missile length	1·87 m
Missile weight (before launch)	75 kg
Missile wing span	650 mm
Guidance system	Semi-automatic command to line of sight
Propulsion	Two stage solid propellant
Velocity	190 m/sec
Range—Maximum	6,000 m
Minimum	800 m
Time of flight to maximum range	32 sec
Warhead	Shaped charge/HEAT/ anti-personnel

The SS 12 has twice the range of SS 11 and much greater lethality. Like SS 11 it can be fired from a ground launcher or from a vehicle. Its semi-automatic command to line of sight guidance system (*Télécommande Automatique*) is the same as that in Harpon and it is much easier to operate than the earlier manual systems. All the layer has to do is to keep the cross wires of the sight on the centre of the target and the missile will then receive commands automatically through its link to correct its flight on to the target. The guidance equipment tracks the missile using an infra-red device, and a small computer calculates the steering commands which are needed to bring the missile on to the line of sight. SS 12 is being replaced by HOT.

Employment

Brazil	France	South Africa
Chile	Iran	Spain
Egypt	Iraq	Turkey
Germany (FRG)	Pakistan	United Arab Emirates

SS 12 ATGW on trailer

HOT anti-tank guided weapon

France and the Federal Republic of Germany

Missile diameter	143 mm
Missile length	1·275 m
Missile weight (before launch)	21 kg
Missile wing span	310 mm
Guidance system	Semi-automatic command to line of sight
Propulsion	Two stage solid propellant
Velocity	260 m/sec
Range—Maximum	4,000 m
Minimum	75 m
Time of flight to maximum range	16 sec
Warhead	Shaped charge

The HOT system (*Haut—Subsonique, Optiquement Téléguidé*) is a collaborative venture by France and West Germany. Its guidance system is similar to that of SS 12 and Harpon, but it is launched from a tube on to the line of sight to the target in the same way as a recoilless gun. This enables it to achieve a very short minimum range. It is normally mounted on armoured vehicles or helicopters.

Although this system of guidance achieves a fast engagement and a short minimum range, it is not compatible with any lateral separation between the operator and the launcher as in Swingfire, Sagger and Swatter.

Employment

Egypt	Iraq	Syria
Germany (FRG)	Kuwait	
France	Saudi Arabia	

The HOT anti-tank missile after launch from AMX 13 tank

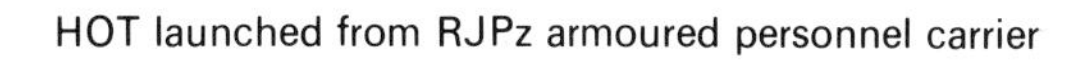

HOT launched from RJPz armoured personnel carrier

TOW anti-tank guided weapon M 151E2

United States

Missile diameter	150 mm
Missile length	1·17 m
Missile weight (before launch)	20·9 kg
Missile wing span	340 mm
Guidance system	Semi-automatic command to line of sight
Propulsion	Two-stage solid propellant
Velocity	Approx 350 m/sec at maximum
Range—Maximum	3,750 m
Minimum	65 m
Time of flight to maximum range	15 sec
Warhead	Shaped charge

The TOW (Tube-launched, Optically-tracked, Wire-guided) system can be mounted on a wide variety of surface and airborne platforms.

The 24 kg missile is fired from a 2·2 m long launching tube which provides initial guidance down the line of sight to the target in the same way as HOT. The semi-automatic guidance system follows the same principles as the HOT, SS 12 and Harpon systems. It is simple to operate since all the operator has to do is to keep the cross wires of the tracker sight on the target. Corrections are transmitted automatically to the missile to bring it on to the line of sight. Again as in HOT fast engagements and a low minimum range are possible, but it has no facility for separation between the operator's sight and the launcher.

Employment

Canada
Denmark
Germany (FRG)
Greece
Iran
Israel
Italy
Jordan
Kenya
South Korea
Kuwait
Lebanon
Luxembourg
Morocco
Netherlands
Norway
Oman
Pakistan
Portugal
Saudi Arabia
Spain
Sweden
Taiwan
Tunisia
Turkey
UK
USA

TOW mounted on a Mule

Swingfire

United Kingdom

Missile length	1·06 m
Missile diameter	170 mm
Missile weight (before launch)	28 kg
Missile wing span	390 mm
Guidance system	Manual command to line of sight
Propulsion	Single stage solid propellant
Velocity	180 m/secs
Range—Maximum	4,000 m
Minimum	150 m (direct mode), 300 m (separated mode)
Time of flight to maximum range	26 sec
Warhead	Shaped charge

Swingfire is mounted on three vehicles in the British Army, the FV 438, Striker and the FV 712 scout car. In the FV 438 it has two launchers which can be replenished from inside the vehicle, whereas on the FV 712 (four launchers) and Striker (five launchers) the missiles have to be replenished from outside.

Swingfire has a most useful facility in that it can be fired either from the periscopic sight on the vehicle or from a separated sight located up to 100 m horizontally and 23 m vertically away from the launcher. This gives great flexibility in siting the launcher in positions where the operator can have the best field of view, whilst keeping the launcher vehicle concealed from direct counter fire from enemy weapons. When the separated sight is used the missile is automatically programmed to fly into the operator's field of view. He then guides it on to the target by manual command signals with his sight thumbstick transmitted by the wire link to the missile.

Swingfire has a longer minimum range and target engagement time than HOT and TOW, but its separation facility gives its operators better flexibility on the battle field. Also its manual command to line of sight guidance system cannot be 'spoofed' by infra-red decoys offset from the target, which are possible counter measures to the semi-automatic systems.

Employment—Belgium, Egypt, Kenya, UK

Swingfire launched from FV 438

Swingfire being fired from Striker

IX. Anti-aircraft Guns and Surface-to-air Missiles

Land tactics have been dominated for the past thirty years by strike aircraft and, other things being equal, success has invariably gone to the side possessing the advantage in the air. Aircraft have always, for obvious reasons, presented a difficult target, but modern aircraft with their high speeds, capability for low-level flight and robustness have led to research and development of ever more sophisticated weapon systems.

To simplify a complex subject, there were two approaches to the problem. One was an accurate solution of the range-speed equation so as to give the gun the correct lead.

The second was, crudely, to fill the sky with projectiles in the hope that one might register a hit and that at least the sight of tracer and bursts might deter all but the most fanatical pilots from following the correct course for bomb-release. The last was uneconomical both in men and ammunition, while the former became ever more difficult as aircraft speeds went up to the neighbourhood of Mach 1.

Larger guns requiring range/height data and prediction were handicapped by imperfections in locating and target acquisition equipment. Prediction was by mechanically driven analogue computers.

Improvements in radar and the development of the electronic digital computer have completely changed this scene, as have the engineering methods which can now produce automated guns without too great a weight penalty.

A number of countries gave up cannon altogether and concentrated on an exclusively guided missile solution. Others retained only what are virtually multiple machine guns of up to 20 mm for purely low level local protection, but the Russians continue to use guns of all types to supplement their excellent inventory of surface to air missiles (SAMs).

SAMs have their limitations. Their control radars are vulnerable to ECM and also to direct action by land or air, as are the missiles themselves when on the ground. Missile supply in the field has its difficulties and it is easy, as is believed to have happened in 1973, to shoot away the entire holding when under heavy air attack. Guns are much easier to resupply and also more versatile tactically. Given well-trained and spirited detachments, anti-aircraft (AA) guns can still be disturbingly effective as the Viet Cong have shown. Whatever air force men may say, in real action pilots have the very greatest respect even for small weapons which theoretically should not be a threat.

There is one other use for the smaller general purpose automatic weapons. The latest weapon carrier and general tactical vehicle on the battlefield is the helicopter, and for this the lighter AA artillery, especially SPs deployed well forward, are a cost-effective antidote.

There are broadly three methods of deploying air defences. The first is point defence. The vital point to be defended could be a bridge, defile or an HQ; the weapons are deployed so as to intercept the aircraft before the point of bomb (or weapon) release is reached. This becomes increasingly difficult if stand-off missiles are available to the attacker.

The second method is area or 'carpet' defence: this is extremely expensive in the number of weapons used for even small areas of importance.

The third is to distribute light AA weapons organically through combat units so as to provide a spread of immediately available low level defence. Tracked SP armoured AA guns with on-carriage radar and fire-control equipment are admirably suited to accompany tank units which themselves are likely to be priority targets for air attack. Hence the ZSU-23-4, ZSU 57-2, M 42, AMX DCA-30 and smaller hand-held missiles e.g. SA-7, SA-9, Redeye and Blowpipe.

Tactical control presents problems. It is not sufficient to hand out guns and leave the rest to the gunners. In open desert or flat tree-less country (where exercises so often are staged), guns can engage over large fields of fire from almost any position. In most terrain the line of sight of guns, sentries and radars are interrupted by the topography, trees, buildings and also the weather. To find good gun positions the countryside will always have to be carefully reconnoitred in advance and deployment controlled so that some weapons are in action and ready to shoot while others are moving to new positions, so ensuring continuous air defence cover.

There is also the question of fire control. Against low level high speed attacks early warning to alert the detachments is essential, and in the stress of battle positive controls must be imposed to ensure the safety of friendly aircraft. In World War II this was also a real problem, but potential more than real, as light air defence was relatively speaking inefficient and was alarming rather than dangerous. Today the hit probability is infinitely higher. The equipments mentioned in this chapter must therefore be seen as only part of a much bigger air defence complex of which guns, SAMs and fighter aircraft are a part, all linked by high speed, sophisticated communications.

The AA gun is a highly specialized weapon which requires elevation to the vertical, all round traverse, a high rate of fire, the ability to fire while the barrel is traversing/elevating and specialised sights. It need not be of a very large calibre, and indeed many are of the machine gun type. In this chapter we are concerned with the heavier equipments which are normally manned by trained artillery personnel.

The problem facing the AA gunner is the small size and great speed of his target. Absolute range is less important than accuracy. The siting of the weapon will be to protect a specific area and since the behaviour of attacking aircraft can be predicted, the individual weapon will be tasked and deployed so as to avoid the random engagement of distant targets.

At short ranges, the gun's zone can be kept within acceptable limits if it is on a suitably stable platform. The choice is invariably a cruciform of legs or girders, an arrangement which requires a four wheel trailer carriage for travelling. The legs can be folded, retracted or hinged and the platform they provide is levelled exactly by jacks. The disadvantages of such equipments are their vulnerability and the time taken to deploy and re-deploy. The alternative for use in the combat zone is the armoured SP gun or multiple gun. However, the SP has its disadvantages; siting must be to give optimum fields of fire, which makes concealment difficult. The smaller, irregularly-shaped, towed gun is easier to deploy and hide. Nor is the need for speedy deployment always paramount. Given that most of the weapons under discussion are not normally associated with the front line, and that they cannot be ubiquitous, the choice of area to defend will probably be slow moving or static; e.g. 'choke-points' on supply routes, H Q, defiles, dumps and logistic service areas, airfields, on which the nature of aircraft attack plans are predictable.

The second factor affecting accuracy is the sight. The movement of the target and the time of flight of the projectile make it necessary to aim off. The amount of aim-off or "lead" varies with the apparent speed of the aircraft movement. An aircraft flying straight at the gun will have no apparent movement relative to the line of sight, and as the angle of approach changes the apparent crossing speed of the target increases until the maximum is reached at a right angle to the line of sight. The simplest sight consists of a conventional fore-sight and a back-sight made up of concentric rings. Its effectiveness depends almost entirely on the skill of the operator—the third factor of accuracy. Tracer ammunition is used to give a constant indication of the operator's performance. Whatever sighting system is used, it relies on a level platform—hence the universal provision of levelling jacks on the trail legs. The whole area of aiming and operating is open to technological assistance. The earliest aid was a mechanical predictor, used mainly with heavy guns of long range. It fed data to the gun via moving pointers, to which the operator matched the lay of the gun—a form of 'indirect' sighting. Today, the electronics industry has evolved radar and computer control systems of great accuracy, and many of these take over the laying of the gun through servo motors. Eldorado (France), Skyguard (Switzerland) and Fire-Can (USSR) are among the equipments available.

A high rate of fire is essential to increase the probability of a hit. Automatic loading from a hopper or magazine is universal for all but the biggest calibre guns. Sliding breech mechanisms, which are usually associated with automatic fire, are used in all the guns described in this section. The recoil is used to activate the fully automatic guns. In many instances, the description of individual guns gives a slower rate of fire than the cyclic rate which is quoted in the tabulated data. The difference is explained by the effort of manually replenishing the hopper or magazine, but this factor is not significant since the average engagement is probably less than a minute.

The projectile need not be large, as the sophisticated equipment on which modern aircraft depend is vulnerable to the slightest damage. It is this fact which leads to the effectiveness of smaller automatic guns with a high rate of fire. A short time of flight and flat trajectory reduce the problems of laying, so a high muzzle velocity is a universal feature. Barrels are long and the propellant charges relatively big but limited in diameter and length by the loading system.

Popular opinion is that the guided missile provides a better air defence weapon than the gun. Certainly the guidance increases its inherent efficiency, but it is not a perfect answer. It is subject to electronic counter measures, minimum ranges and technical failure, and it is enormously expensive; properly handled guns can also be effective. The efficiency of North Vietnamese 'conventional' AA fire provided a valuable lesson. The answer in the immediate future is a balanced mixture of guns and missiles and, to look at the subject from a wider perspective, a whole spectrum of fire-power ranging from infantry small arms fire, missiles and small automatic cannon handled by non-specialists to guns and major missiles operated by specialist artillery personnel; to which must be added, of course, air-to-air fire-power, all controlled or supported centrally, as far as possible and connected by an elaborate system of surveillance and telecommunications.

Twin 20 mm M3/VDA

France

Calibre	20 mm
Barrel length	2,600 mm
Shell weight	125 g
Cyclic rate of fire	2 × 1,000 rds/min
Muzzle velocity	1,050 m/sec
Maximum horizontal range	2,000 m
Maximum effective slant range	1,500 m
Ceiling	800 m
Detachment	3
Chassis—Type	Panhard M 3/VTT
Engine	1997 cc Panhard 4 HD
Power	90 bhp at 4,700 rev/min
Maximum speed	100 km/hr
Endurance	1,000 km
Ammunition carried	650
Height × length × width	2·1 × 3·79 × 2·0 m
Weight	6,300 kg

The twin 20 mm HS 820 SL guns are mounted on a one-man turret on a Panhard M 3 scoutcar. It is a highly mobile weapon system with speeds of up to 100 km/hr possible over a range of 1,000 km. The EMD X band pulse doppler radar has an acquisition range of 8 km. When a target has been acquired, the bearing and range are automatically fed to the sight by the P 56 Galileo computer. All the gunner has to do is to keep the sight on the target and the computer automatically lays off the correct lead and range, and indicates the moment to open fire. It is thus particularly suitable for short range air defence in support of fast moving operations, although the number of ready to use shells it can carry (650 in two 325 round belts) could be a serious limitation.

Employment France

Twin 20 mm M3/VDA on Panhard scout car

Twin 20 mm RH 202 Mk 20

Federal Republic of Germany

Calibre	20 mm
Barrel length	1·84 m
Weight	2,160 kg
Shell weight	HE-IT, APT } linked 250 gm
Cyclic rate of fire	2,000 rds/min
Muzzle velocity	1,050 m/sec
Maximum horizontal range	2,000 m
Maximum slant range	1,800 m
Ceiling	900 m
Detachment	3
Ammunition carried	550
Height × length × width	2·07 × 5·03 × 2·36 m

The Rheinmetall 202 Mk 20 is a twin 20 mm gun system mounted on a lightweight two wheeled trailer. The sighting system is a simple optical device, incorporating rigid flight direction lines and a variable prediction line in elevation. It is thus a daylight only weapon. The gunner has to estimate the target speed and range to his intended point of engagement and then to set it on the lead computer. As soon as tracking begins, information is automatically fed into a simple computer, which in turn adjusts the lead marks in the monocular sight. Power for the hydrostatic drives is provided by a petrol driven 8 hp Wankel generator mounted behind the layer's seat. 550 rounds in two belts of 275 are carried on the gun. The gun is towed by a $1\frac{3}{4}$ ton truck and it is operated by three men—a gunner and two ammunition handlers.

Employment Germany (FRG)
Norway

The RH 202 Mk 20 in action

20 mm Vulcan (M 163)

United States of America

Calibre	20 mm
Barrel length	—
Shell weight	M 50 HE IT-SD, HE 1, TP-T, TP } 101 gm
Cyclic rate of fire	3,000 rds/min
Muzzle velocity	1,088 m/sec
Maximum horizontal range	3,000 m
Maximum effective slant range	2,000 m
Ceiling	900 m
Detachment	4
Chassis—Type	M 113 A 1
Engine	GMC Model 6 V 53 diesel
Power	215 hp
Maximum speed	68 km/hour
Endurance	480 km
Ammunition carried	1,900
Height × length × width	2·74 × 4·87 × 2·81 m
Weight	11,208 kg

The Vulcan gun on M 113 chassis

This six barrelled 20 mm gatling gun is mounted on the M 113A1 APC chassis. Its ammunition feed system does not have links, which reduces the chance of stoppages in its very high rate of fire guns. The X band pulse-doppler radar acquires targets out to 5,000 m, but it only indicates range and bearing, target tracking being visual through a lead-computing sight. Although the six barrels have an excellent rate of fire the number of ready to use rounds carried is a severe limitation and it means that each gun must be accompanied by its own ammunition supply vehicle. In the US Army local warning is passed to Vulcan by the Forward Area Alerting Radar (FAAR) system.

The M 167 is a much lighter wheeled trailer version (1,500 kg) of Vulcan which can be towed by a $1\frac{1}{4}$ ton truck. Both versions have a detachment of four men—commander, layer, ammunition number and driver.

Employment Israel
USA

The Vulcan gun M 167, trailer mounted

23 mm SP anti-aircraft gun ZSU-23-4 'Shilka'

Union of Soviet Socialist Republics

Calibre	23 mm
Barrel length	82 calibres
Ammunition options	HEI-T (188·5 gm) API-T (190 gm)
Cyclic rate of fire	1,000 rds/min/gun
Muzzle velocity	970 m/sec
Maximum horizontal range	2,200 m
Maximum effective slant range	2,000 m
Ceiling	1,000 m
Detachment	4
Chassis—Type	PT 76
Engine	V-6, 6 cylinder diesel
Power	240 hp at 1,800 rev/min
Maximum speed	44 km/hour
Endurance	260 km
Ammunition carried	2,000 plus 3,000 in support vehicle
Height × length × width	2·25 × 6·10 × 2·80 m
Weight	14,000 kg

ZSU-23-4 'Shilka' in action

This gun was first introduced into the Soviet Army in 1965 using the PT 76/ASU-85 chassis. It has four 23 mm guns each able to fire at 1,000 rounds per minute. The upper mounting has 360° traverse and allows the guns to elevate to 85°. Fire control is either by optical sights, which are housed on the turret behind the guns, or by the 'Gundish' radar which has an acquisition range of 8,000 m. This was the most effective gun in the October 1973 war, accounting for a high percentage of Israeli aircraft losses. With the SA-9 missile system it provides a comprehensive very low/low level air defence of forward areas. Many Soviet-equipped countries have a twin 23 mm version of the same gun mounted on a trailer.

Employment

Bulgaria	Germany (GDR)	Poland
Czechoslovakia	Hungary	Syria
Egypt	India	USSR
Finland	Iran	

23 mm SP AA gun ZSU-23-4

30 mm SP anti-aircraft gun 'Falcon'

United Kingdom

Calibre	30 mm
Barrel length	—
Shell weight	0·36 kg
Cyclic rate of fire	650 rds/barrel/min
Muzzle velocity	1,080 m/sec
Maximum horizontal range	10,000 m
Maximum effective slant range	3,000 m
Ceiling	1,000 m
Detachment	3
Chassis—Type	Abbot SP gun
Engine	Rolls Royce K60
Power	213 bhp
Speed	48 km/hour
Endurance	390 km
Ammunition stowage	620 rds
Height × length × width	2·51 × 5·3 × 2·6 m
Weight	15,850 kg

The Falcon has been developed by Vickers Ltd. and consists of two HSS 831L 30 mm cannon mounted in a turret on the Abbot chassis. Both components have been described elsewhere but can be summarised as follows.

The chassis is that of the Abbot SP gun and is based on the FV 430 series Armoured Personnel Carrier. It runs on five road wheels with engine and driver at the front. The Rolls Royce engine is a six cylinder horizontally opposed two-stroke turbo-charged diesel. The transmission is by Alison automatic gear box and Cletrac steering unit. The chassis has an excellent and proven cross-country performance.

The guns are made by Oerlikon-Buhrle in Switzerland. Their basic design allows a number of ammunition feed options, and in this application the open link belt is used feeding from the inboard side of each gun. The barrels are bayonet fitted to the body for quick changing.

The mounting is in a new purpose-built turret which is armour-plated against small arms fire and shell splinters. The commander's and gunner's seats are adjustable; access to them is through the roof. The fighting compartment is sealed from the guns to avoid fumes fouling the atmosphere, and to reduce noise when firing. The guns are cocked and fired (single shot or automatic) electrically.

The boxed ammunition (HE, AP or practice) is stored in the chassis in a compartment sealed from the turret. The ammunition belts feed to the turret by chutes. Spent cartridges and belt links are ejected sideways through the elevation trunnion bearings.

Elevation and traverse are electrically powered and controlled by a joystick. The commander has his own control which overrides the gunner's. The guns are stabilised for firing on the move by a rate gyro system. This system, made by GEC-AEI (Electronics) Ltd., is based on that used in the Chieftain tank.

The fire control system is optical, supported by a small predictor computer. The gun is invulnerable to electronic counter measures, cheap to build and easy to maintain because it has no radar. The gunner's periscopic sight has a 50° field of view with a circular aiming mark displayed in the viewer. The aiming display is computer controlled to provide automatic compensation for target movement. The range is set manually, and is used by the computer to determine the optimum lead by reference to the rate of tracking. A laser rangefinder can now be incorporated in the fire control system.

A ×6 magnification periscopic sight is also provided for use against ground targets.

Employment Development only in UK.

Falcon gun on Abbot chassis

Twin 30 mm anti-aircraft gun M 53

Czechoslovakia

Calibre	30 mm
Barrel length	70 calibres
Weight	2,100 kg
Shell weight	HE (450 gm) API (450 gm)
Cyclic rate of fire	500 rds/min/barrel
Muzzle velocity	1,000 m/sec
Maximum horizontal range	10,000 m
Maximum effective slant range	3,000 m
Ceiling	1,000 m
Armour penetration	API 60 mm at 500 m
Detachment	4

This twin barrelled gun was first seen in 1958, and remains in service with the Czech divisional anti-aircraft regiments. The very long barrels with their multi-baffle muzzle brakes are fitted in quick-change mountings. Ammunition in 10 round clips, is loaded into a horizontal feed on the left of the gun. This system limits the rate of fire to 100 rounds/minute/barrel, although the cyclic rate for fire is much higher.

There are two types of ammunition: HE and AP. Both weigh 0·45 kg, and give the same performance. The AP round will penetrate 60 mm of armour at 500 m. The gun has a modest range, and relies on optical laying using a cart wheel sight. Traverse and elevation are hydraulically powered.

The guns are mounted on a light, four-wheel trailer without any protective armour. The weapon fires on its wheels with screw-type firing jacks beneath each axle. An SP version, the M 53/59 is described separately.

Employment	Cuba	Iraq
	Czechoslovakia	Uganda
	India	

Twin 30 mm M 53

30 mm Type GCI anti-aircraft cannon

Switzerland

Calibre	30 mm
Barrel length	75 calibres
Weight	1,550 kg
Shell weight	0·36 kg
Cyclic rate of fire	600–650 rds/min
Muzzle velocity	1,080 m/sec
Maximum horizontal range	—
Maximum effective slant range	3,000 m
Ceiling	1,000 m
Detachment	1

The type GCI consists of the Oerlikon KCB automatic cannon on a simple trailer mounting, and is highly mobile. It has a small, circular platform, with four outrigger stabilizers which fold for travelling. The wheels are removed for firing.

The gun is operated and fired by one man using a joystick to operate hydraulic control in elevation and azimuth, and a Type P 36 Galileo mechanical fire control computer sight. The layer sits behind the gun astride the rear mounting, and has some body protection from a contoured shield. Hydraulic power is supplied from a pump driven by a small Wankel engine.

For this mounting, the magazine loading system, feeding from the left is used.

Employment Switzerland

30 mm GCI in action. (Formerly known as HS 661)

Twin 30 mm SP anti-aircraft gun M 53/59

Czechoslovakia

Calibre	30 mm
Barrel length	70 calibres
Shell weight	0·45 kg (HE and API)
Cyclic rate of fire	500 rds/min/barrel
Muzzle velocity	1,000 m/sec
Maximum horizontal range	10,000 m
Maximum effective slant range	7,000 m
Ceiling	2,000 m
Detachment	5
Chassis—Type	PRAGA V3S 6 × 6
Engine	Tatra 6-cylinder diesel
Power	110 bhp
Maximum speed	60 km/hr
Endurance	500 km
Ammunition carried	200
Height × length × width	3·06 × 6·98 × 2·41 m
Weight	9,500 kg

This is the SP version of the M 53 with twin guns in a partially enclosed mounting installed on the armoured version of the PRAGA V3S 6 × 6. The multi-baffle muzzle brake and hydraulic control system are identical with the M 53 but the loading system is altered to a magazine using a vertical feed. The rectangular curved magazine holds 50 rounds and aids the recognition of this version. Two spare magazines are carried on the gun tray of the vehicle.

This weapon is in service with divisional anti-aircraft regiments and the anti-aircraft companies of the armoured regiments of the Czech Army.

Employment Cuba
Czechoslovakia
Yugoslavia

The SP M 53/59 on parade

---

30 mm twin SP AMX-DCA 30

France

Calibre	30 mm
Barrel length	—
Shell weight	0·36 kg
Cyclic rate of fire	600 rds/min
Muzzle velocity	1,080 m/sec
Maximum horizontal range	10,000 m
Maximum effective slant range	3,000 m
Ceiling	1,000 m
Detachment	4
Chassis—Type	AMX 30
Engine	Hispano Suiza HS110
Power	720 bhp
Maximum speed	65 km/hour
Endurance	—
Ammunition carried	—
Height × length × width	3 × 6·8 × 3·1 m
Weight	36,000 kg

This is the HO-831 gun, twin mounted in the SAMM S40 1A turret on an AMX 30 tank chassis. The turret and guns were introduced in 1964 mounted on the lighter AMX 13 tank chassis. The new chassis, which is now coming into service, carries more ammunition, and has an electric power supply which is independent of the main engine. With this mounting and the introduction of an integral fire control radar (the Oeil Noir) the weapon becomes self-contained. In order to be most effective it must have early warning from an area radar system, but the guns can then lock to the on-board radar data, using computer control for the calculation of essential lead angles. The optical sights are for emergency use only.

The AMX 30 is a large vehicle: the SAMM turret adds 0·71 m to its height even when the radar dish is folded flat for travelling. Despite its weight (about 36,000 kilogrammes) the flat 12 cylinder multi-fuel super-charged engine gives 20 bhp/ton, and gives a good performance.

Employment France

The SP AMX-DCA 30 with guns elevated and radar dish (*behind*) erect for action

---

35 mm SP anti-aircraft gun 'Gepard'

Federal Republic of Germany

Calibre	35 mm
Barrel length	90 calibres
Shell weight	0·55 kg
Cyclic rate of fire	550 rds/min/barrel
Muzzle velocity	1,175 m/sec
Maximum horizontal range	12,000 m
Maximum effective range	4,000 m
Ceiling	1,200 m
Detachment	3
Chassis—Type	Leopard main Battle tank
Engine	Daimler-Benz multi-fuel
Power	830 bhp
Maximum speed	65 km/hour
Endurance	600 km
Ammunition stowage	700
Height × length × width	3·0 × 6·94 × 3·25 m
Weight	45,100 kg

The Gepard, or Flakpanzer 1, has been designed to support forward combat teams. It consists of two

Oerlikon 35 mm cannon mounted in a fully enclosed turret on the Leopard main battle tank chassis. The vehicle incorporates tracking and search radar, and provides its own electrical power. It is therefore a fully independent weapon system. The first prototype was completed in 1968.

The guns are those of the Swiss GDF-001, but are modified to belt feed. The cyclic rate of fire is not therefore limited by the need for manual loading of clips as it is with the original equipment. However, the on-board ammunition supply is very limited. It is probable that the radar and fire control equipment take up much of the space in the fighting compartment—but these very arrangements allow best use to be made of ammunition in short, accurate engagements.

Both HE and AP rounds are available. Lethality is claimed to equal the 40 mm Bofors shell, and the AP round penetrates 44 mm of armour at 1,000 m.

The guns are mounted either side of the turret, the tracking radar head between them at the front, and the search radar antenna above and to the rear. The search radar is the Siemens MPDR-12. It is a pulse doppler type, incorporates IFF and is fitted with an automatic alarm. Transfer to the tracking radar is automatic on the operators command. The mounting of the tracking radar allows targets in a 200° arc to be accepted without slewing the turret. It is a monopulse doppler type made by Siemens-Albis. Target data for the fire control computer which calculates the optimum point of aim is augmented by muzzle velocity data from the guns: the distinctive measuring devices fitted to the muzzles are the same as those used on the K 63.

All electronics include a self test facility. The fire control computer is covered by a small emergency computer, and the whole automatic system is backed-up by an optical system. This system is periscopic for best crew protection, and allows for visual target acquisition, battlefield surveillance and gun sighting against ground and air targets. To allow for firing on the move the sights are gyro stabilised.

The Leopard chassis dates from 1965. It has good battlefield mobility which stems from a power to weight ratio of 21 bhp/ton. The powerful engine is a supercharged, compression ignition, multifuel V 10. Drive is transmitted through a four speed automatic gearbox. The vehicle is fully proofed against nuclear radiation and chemical attack, and wades to a depth of 1·2 m without special preparation. In the Gepard version a 90 bhp diesel engine powers a 60 kilowatt generator.

Although Gepard appeared first as a German Army

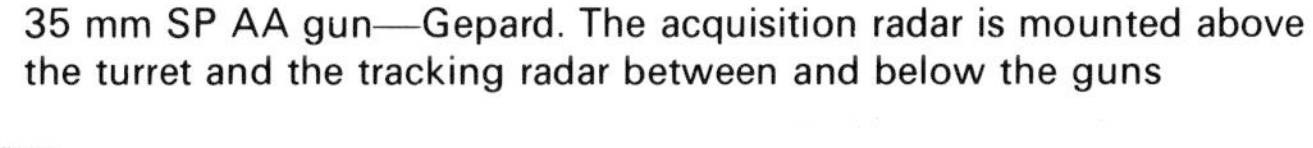

35 mm SP AA gun—Gepard. The acquisition radar is mounted above the turret and the tracking radar between and below the guns

equipment, it is largely the result of international co-operation. The first prototypes were built by Contraves AG of Switzerland, using a German chassis which is derived from the Leopard Main Battle Tank. In the production models the chassis is built by Krauss-Maffei of Munich; the guns are by Oerlikon-Buhrle AG of Zurich; the search radar by Siemens AG Munich; the tracking radar by Siemens-Albis, Zurich, and the sights and fire control computer by Contraves AG of Zurich.

A second version developed for the Netherlands Army retains the same guns and fire control but uses a combined search and track radar made by NV Signaalapparaten and the assembly is done in Holland.

Employment Belgium
Germany (FRG)
Italy
Netherlands

Twin 35 mm anti-aircraft gun K 63

Switzerland

Calibre	35 mm
Barrel length	90 calibres
Weight	6,400 kg
Shell weight	0·55 kg
Cyclic rate of fire	550 rds/min/barrel
Muzzle velocity	1,175 m/sec
Maximum horizontal range	—
Maximum effective slant range	4,000 m
Ceiling	1,200 m
Detachment	4
Ammunition carried	238

This Oerlikon gun is used in both this, land based mounting or as a ship-borne anti-aircraft gun. It is known in Austria as the M 65 and has also been built under licence in Japan. The high cyclic rate of fire cannot be sustained in practice because of the manual feed system using six round clips. Both HE and AP ammunition is available: the AP round penetrates 44 mm of armour at 1,000 m.

The gun is constructed so that the barrels and their feed systems are interchangeable between left and right mounting.

The four road wheels are swung clear of the ground for firing but remain on their axles. Two outriggers make up a cruciform platform with hydraulic jacks at each end of the chassis, and on each outrigger: time into action—1½ minutes.

Three modes of operation are provided: fully automated electrical control linked to Super Fledermaus radar: by column control with optical sights: by mechanical hand wheels. The weapon is fitted with a device which constantly monitors muzzle velocity. The very high laying speeds achieved by the electrical motors are a feature of the design: 360° traverse is achieved in three seconds.

Employment Austria, Finland, Japan, South Africa, Switzerland

The K 63 in action with the road wheels folded on their axles over the carriage, the outriggers deployed and the whole weapon supported on hydraulic jacks. Note the muzzle velocity measuring device between the barrels

37 mm anti-aircraft gun M 1939

Union of Soviet Socialist Republics

Calibre	37 mm
Barrel length	70 calibres
Weight	2,050 kg
Shell weight	0·74 kg (HE) 0·77 kg (AP) 0·62 kg (HVAP)
Cyclic rate of fire	180 rds/min
Muzzle velocity	960 m/sec (HVAP) 880 m/sec (HE and AP)
Maximum horizontal range	8,000 m
Maximum effective slant range	3,000 m
Ceiling	1,000 m
Detachment	8

This gun is very similar to the Bofors L/60 from which it derives. The conical flash eliminator is one detail which is retained from the original design. Loading is manual: clips of five rounds are loaded into a hopper with gravity feed to the vertically opening sliding breech. Extraction of spent cartridges is automatic down a chute beneath the breech. The cylic rate of fire is only 160 rounds/minute which compares unfavourably with the rate of the 40 mm Bofors gun L/70. The manual loading arrangements further reduce the rate in practice to about 80 rounds/minute. The armour penetration of the AP shell is 46 mm at 500 m.

Sighting is optical. Split functions laying is used and is manual.

The gun is mounted on a four-wheel carriage. For action the wheels are removed, and a screw type jack supports each axle. Two swinging outriggers are deployed to complete a cruciform firing platform. A shield is sometimes fitted.

The M 1939s of the Soviet Army have been relegated to the reserve. The Chinese guns of this type are known as M 55.

Employment:

Afghanistan	Egypt	Somalia
Albania	Germany (GDR)	Sudan
Algeria	Kampuchea	Tanzania
Bulgaria	North Korea	USSR
China	Laos	Vietnam
Congo	Mongolia	North Yemen
Cuba	Morocco	Yugoslavia
Czechoslovakia	Pakistan	Zaire

Bringing the gun into action

Below: The gun action, legs extended, jacks lowered and ammunition hopper in position

40 mm anti-aircraft gun L/60

Sweden

Calibre	40 mm
Barrel length	60 calibres
Weight	1,750 kg
Shell weight	0·96 kg
Cyclic rate of fire	120 rds/min
Muzzle velocity	850 m/sec
Maximum horizontal range	8,700 m
Maximum effective slant range	3,000 m
Ceiling	1,000 m
Detachment	7

This pre-World War II equipment has been produced in a number of versions with many different control and sighting systems. It is mounted on a four wheel carriage, and may be fired from its wheels, although the wheels are normally removed for firing and stability is obtained by using two girder outriggers and four jacks. Ammunition is hand loaded in clips of four rounds to a hopper with a vertical feed. There is automatic extraction of the empty case through a chute mounted below the gun. There is a flat shield to provide protection to the crew, and this together with the distinctive conical flash eliminator form useful recognition features. AP rounds are available, in addition to HE, to give the gun a surface-to-surface capability. Armour penetration is 55 mm at 500 m.

40 mm Bofors L/60 AA gun

Employment

Argentina, Austria, Brazil, Burma, Cambodia, Cyprus, Ecuador, Egypt, Eire, Finland, Greece, India, Indonesia, Israel, Ivory Coast, Jordan, Lebanon, Libya, Malaysia, Nepal, Nigeria, Norway, Pakistan, Sudan, Taiwan, Thailand, Tunisia, Turkey, Yugoslavia, Zaire

40 mm automatic gun L/70

Sweden

Calibre	40 mm
Barrel length	70 calibres
Weight	5,150 kg
Shell weight	1 kg
Cyclic rate of fire	240 rds/min
Muzzle velocity	1,000 m/sec
Maximum horizontal range	12,500 m
Maximum effective slant range	3,000 m
Ceiling	1,200 m
Detachment	7

This gun was developed as a successor to the L/60 and takes its title from the calibre length of the barrel. It first came into service in 1951 and has become widely used since that time. Over 5,000 have been built.

Ammunition in clips of four rounds is hand fed by two loading numbers to a hopper with a vertical feed. It is possible to stack up to 16 rounds in the hopper for ready use, and a further 48 rounds are carried at the rear of the platform. HE, AP and APDS ammunition is available. The HE round is fitted with a percussion fuze, and a self-destruct fuze set for 8·5 seconds after firing. A new and reputedly highly effective pre-fragmented round, with proximity fuze has been developed.

The gun is mounted on a four wheel trailer and has a good cross-country capability; for firing the wheels are normally removed and two swinging outriggers provide a cruciform for stability, the gun being levelled by means of four jacks. The gun is usually seen with a wrap-round armoured shield, and its principal recognition feature is the long slender barrel with a conical flash eliminator. Some models have built-in power units, but the weapon usually relies on an external power source.

Control is normally by a single layer using a control column and an electro-hydraulic elevating and traversing system, but provision is made for hand control using two hand wheels and mechanical drive. Provision is also made for remote control from any of the associated fire control radars using a cable link. A number of radar control systems are used; the principle types are:

Eldorado-Mirado (France) L/45 (Netherlands)
Super Fledermaus (Switzerland)
PS 04R (Sweden)

An Italian version of this gun is known as the Breda/Bofors. It can be fitted with an automatic feed working from a 144 round magazine.

Employment	Australia	India	Japan
	Austria	Iran	Netherlands
	Canada	Israel	Norway
	France	Italy	Spain
			Sweden

40 mm Bofors L/70 AA gun

Bofors 40 mm 75

Sweden

Calibre	40 mm
Barrel length	70 calibre
Weight	5,150 kg
Shell weight	Prefragmented proximity fuzed (875 g) HE tracer (960 g) AP tracer (930 g)
Cyclic rate of fire	300 rds/min
Muzzle velocity	1,035 m/secs
Maximum horizontal range	12,500 m 4,000 m
Maximum effective slant range	3,500 m
Ceiling	1,250 m
Detachment	5

The Bofors gun system 75 is a modification of the L 40/70, and it incorporates the same 40 mm gun. However it represents a considerable improvement over the L 40/70, and Bofors claim that it is about ten times more effective. It has a miniaturised fire control system and a power supply unit mounted on it, and thus it only requires one towing vehicle.

The fire control system, called BOFI (Bofors Optronic Fire-control Instrument), consists of a day and night sight, a range finding laser and a computer which automatically calculates the aim-off position. The optical target indicator enables the commander standing to the side of the gun, to give an automatic indication of an enemy aircraft to the layer. A petrol-driven four stroke generator is mounted at the rear of the lower carriage and provides the necessary power to the electrohydraulic laying mechanism and to the fire control equipment. In addition to the new proximity-fuzed prefragmented ammunition, the 40/75 can fire conventional high explosive, armour-piercing or practice rounds. A total of 122 ready-to-use shells can be carried on the gun—26 in the hopper and 96 in the racks. It is towed by an 8 tonne truck or equivalent and it has a detachment of five—commander, layer, two loaders and ammunition handler/driver.

Employment Sweden

The Bofors 40 mm 75 AA gun

40 mm twin SP M 42

United States of America

Calibre	40 mm
Barrel length	60 calibres
Shell weight	0·96 kg
Cyclic rate of fire	120 rds/min/barrel
Muzzle velocity	850 m/sec
Maximum horizontal range	8,700 m
Maximum effective slant range	3,200 m
Ceiling	1,000 m
Detachment	4
Chassis—Type	Full tracked
Engine	6-cyl supercharged petrol
Power	500 bhp
Maximum speed	72 km/hour
Endurance	160 km
Ammunition carried	—
Height × length × width	2·84 × 6·35 × 3·2 m
Weight	22,500 kg

This is a fully tracked armoured chassis mounting twin L/60 Bofors guns and has been in service since 1953. The guns are mounted in a cylindrical open-topped turret which is operated under power control (although hand operation is possible). As with the L/60 the gun is hand loaded with clips of four rounds to hoppers providing vertical feed. The sighting system incorporates a computer for the calculation of aim off but there is no radar capability and visual aim is required which limits the range and operational effectiveness. Although the M 42 is heavy it does not afford much protection to its crew, and due to its lack of radar it does not meet the modern requirement very well. It is being replaced by Vulcan and Chaparral in the US Army.

Employment

Austria	Japan	USA (Reserve)
Germany (FRG)	Jordan	
Italy	Lebanon	

40 mm twin SP M 42 AA gun

57 mm anti-aircraft gun

Czechoslovakia

Calibre	57 mm
Barrel length	60 calibres
Weight	5,150 kg
Shell weight	2.5 kg
Cyclic rate of fire	100 rds/min
Muzzle velocity	1,000 m/sec
Maximum horizontal range	12,000 m
Maximum effective slant range	4,000 m
Ceiling	2,000 m
Detachment	7

This gun was produced in the Skoda works in very limited numbers, and is similar to the Russian S 60 which eventually replaced it. The gun is slightly heavier than the S 60 and has a lower muzzle velocity. It is mounted on a four wheel carriage which is lowered to bring the gun to the firing position, but the wheels are not removed. Laying is manual power assisted.

These guns were sold by the Czechs when they accepted S 60 into service.

Employment Cuba
Guinea
Mali

57 mm AA gun—similar to D-60

57 mm anti-aircraft gun M 54

Sweden

Calibre	57 mm
Barrel length	60 calibres
Weight	8,100 kg
Shell weight	2·6 kg
Cyclic rate of fire	120 rds/min
Muzzle velocity	920 m/sec
Maximum horizontal range	14,500 m
Maximum effective slant range	4,000 m
Ceiling	2,000 m
Detachment	6+

The M 54 is closely related to the Bofors 40 mm weapon. The layout and operation of the gun follows the pattern of the L/70, and the typically Bofors conical flash eliminator is fitted. The cyclic rate of fire is high for the calibre of the weapon, but this rate cannot be maintained in practice because of the problem of handling the ammunition. The ready use ammunition racks used on the L/70 are retained.

The gun is mounted on a four-wheel carriage. In action it is supported on four jacks, which serve to level the equipment. Two of the screw-type jacks lift the wheels clear from the ground, the other two are mounted on swinging outriggers which make up a cruciform mounting. The large shield at the front and sides of the gun is an important recognition feature.

Employment Belgium
Sweden

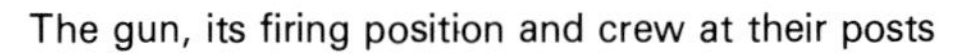

The gun, its firing position and crew at their posts

57 mm anti-aircraft gun M 1950 (S 60)

Union of Soviet Socialist Republics

Calibre	57 mm
Barrel length	73 calibres
Weight	4,000 kg
Shell weight	2·81 kg (HE) 2·83 kg (AP)
Armour penetration	APHE 86 mm at 500 m HVAP 100 mm at 500 m
Cyclic rate of fire	100–120 rds/min
Muzzle velocity	1,000 m/sec
Maximum horizontal range	15,000 m
Maximum effective slant range	4,000 m
Ceiling	2,000 m
Detachment	8

This equipment, which was first seen in 1950, uses the same gun as the ZSU 57-2 SP anti-aircraft equipment, and the ASU 57, M 1943 and M 1955 anti-tank guns. It was developed by the Russians from a captured German gun which was in its final development stages.

Ammunition is loaded into a slide feeding mechanism in clips of four rounds: this handling process reduces the rate of fire to 70 rounds/minute. A proximity fuze is used. The APHE round penetrates 106 mm of armour at 500 m. The gun is mounted on a four wheel carriage. It can fire on its wheels, but to an effective slant range of only 4,000 m, Normally, two swinging outriggers are deployed in action, and the carriage is supported on four levelling jacks. The wheels are not removed in action: they are fixed to trailing-arm suspension members, which, when raised to the horizontal, lift the wheels clean off the ground. For travelling the suspension arms are at a distinctive 45° angle. The pepper-pot muzzle brake is an equally distinctive recognition feature.

Three fire control options are available: a computing sight is fitted to the weapon, the SON 9 radar in conjunction with the PUAZO 6/60 director, and the Firecan S-band radar fire control system. The SON 9 is being replaced by SON 9A mounted on the URAL 375, and known as Flapwheel.

Employment:

Afghanistan, Albania, Bulgaria, China, Congo, Czechoslovakia, Egypt, Germany (GDR), Indonesia, Iran, Iraq, Kampuchea, North Korea, Libya, Mongolia, Pakistan, Poland, Rumania, Syria, Vietnam, South Yemen, Yugoslavia

57 mm AA gun M 1950 (S 60)

57 mm SP ZSU 57-2

Union of Soviet Socialist Republics

Calibre	57 mm
Barrel length	73 calibres
Shell weight	2·8 kg (HE) 3·1 kg (AP)
Cyclic rate of fire	210–240 rds/min/barrel
Muzzle velocity	1,000 m/sec
Maximum horizontal range	15,000 m
Maximum effective slant range	4,000 m
Ceiling	2,000 m
Detachment	6
Chassis—Type	Modified T 54 tank
Engine	V-12 diesel
Power	520 bhp
Maximum speed	48 km/hour
Endurance	400 km
Height × length × width	2·75 × 8·48 × 3·27 m
Weight	28,100 kg
Ammunition carried	316

This is the SP version of the S 60 mounting two guns in a distinctive square turret above a flat topped tracked chassis with a drivers compartment in the extreme left front corner. The recognition features in addition to the distinctive turret shape are the pepperpot muzzle brakes. The turret is open topped which limits the protection offered to the detachment.

Optical sights are fitted and the equipment is thus limited to fair weather and daylight operation.

Employment Czechoslovakia, Egypt, Finland, Hungary, Iran, North Korea, Poland, USSR, Vietnam, Yugoslavia

The 57 mm SP ZSU 57-2

75 mm anti-aircraft gun M 51 (Skysweeper)

United States of America

Calibre	75 mm
Barrel length	40 calibres
Shell weight	6·8 kg (HE)
Cyclic rate of fire	45 rds/min
Muzzle velocity	853 m/sec
Maximum horizontal range	13,180 m
Maximum effective slant range	11,200 m
Ceiling	9,150 m
Detachment	6

Development of this gun began in September 1944 with the intention of taking advantage of the proximity fuze to produce a fast-firing weapon in which there would be no delay due to fuze-setting. At the same time, advances in radar technology and electronics promised the ability to make the radar and fire control instrumentation an integral part of the gun mounting so that an individual weapon could act in a detached role. The realisation of such an advanced concept took several years and it was not until the early 1950s that the gun finally entered service.

The mounting, of the usual cruciform type carried on four removeable wheels, carries the gun, its auto-loading mechanism, an optical tracker, range radar, and fire control computer. Two revolver-type cylinders at the rear hold ten rounds each and feed from alternate sides to a central rammer unit.

Although no longer in use by the US Army, numbers are retained in service by other nations and the gun is said to have a useful secondary function as a fast ground bombardment weapon.

Employment Greece Japan Turkey

75 mm AA gun M 51 (Skysweeper)

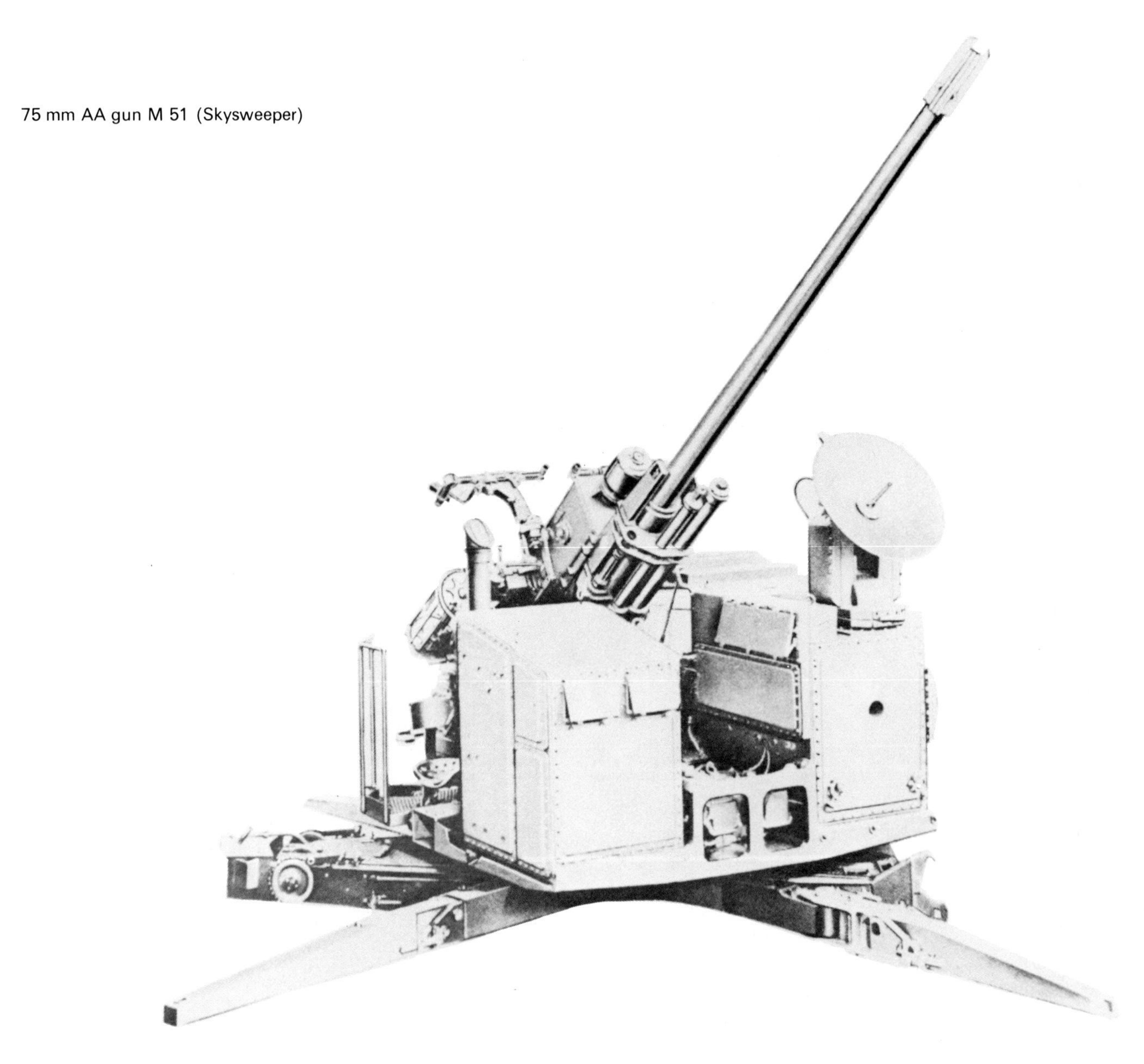

85 mm anti-aircraft gun M 1944 (KS 18)

Union of Soviet Socialist Republics

Calibre	85 mm
Barrel length	53 calibres
Weight	4,300 kg
Shell weight	9·75 kg (HE)
	9·3 kg (APHE)
	5·0 kg (HVAP)
Cyclic rate of fire	15–20 rds/min
Muzzle velocity	1,030 m/sec (HVAP)
	792 m/sec (HE and APHE)
Maximum horizontal range	15,500 m
Maximum effective slant range	9,400 m
Ceiling	3,000 m
Detachment	7

The KS 18 is a development of the pre-World War II KS 12, which is now obsolete. The gun is also used as an anti-tank and tank gun.

The KS 18 does not have an automatic loading system. The rate of fire is therefore poor by anti-aircraft standards, but is helped by a semi-automatic vertically sliding breech, and fixed ammunition. The hydro-pneumatic recoil gear is mounted under and over the barrel. A singular conical multi-baffle muzzle brake is fitted.

The gun is pedestal mounted on a circular platform on a four wheel carriage. Two swinging outriggers make-up a cruciform mounting in action: the wheels are removed.

Both anti-aircraft and anti-tank optical sights are fitted, giving the gun an important dual capability. Fire control by SON 9 Radar and PUAZO 6/12 director is also available, and S 60 Firecan radar control is compatible with this gun.

The Czech Army is equipped with a similar gun manufactured by themselves which is usually known as the M 53 (not to be confused with the 30 mm M 53). This gun is identical in almost all respects, differences being a T-shaped muzzle brake and minor modifications to the shield and travelling platform.

Employment

Afghanistan	East Germany	Sudan
Albania	Hungary	South Yemen
Bulgaria	Iran	Syria
China	Iraq	Vietnam
Cuba	Kampuchea	Yugoslavia
Czechoslovakia	North Korea	
Egypt	Poland	

A section of 85 mm AA M 1944 KS (18) on the march

90 mm anti-aircraft Gun M 118

United States of America

Calibre	90 mm
Barrel length	50 calibres
Weight	14,650 kg
Shell weight	10·6 kg (HE) 10·9 kg (AP)
Rate of fire	27 rds/min
Muzzle velocity	823 m/sec
Maximum horizontal range	17,850 m
Maximum effective slant range	15,200 m
Ceiling	12,040 m
Detachment	9

The 90 mm gun was developed in the late 1930s and standardised in March 1940. It was later improved by the adoption of the M 2 mounting, a four-wheeled cruciform type which allowed the gun to be fired from its wheels and which was also suited to deployment as a harbour defence weapon. It also had a useful anti-tank capability and the gun formed the starting point for the development of a series of tank and anti-tank guns.

The M 2 mount incorporates a highly efficient fuze setter/rammer combination which was instrumental in bringing the rate of fire up to 27 rounds/minute from the original 15.

The 90 mm gun was phased out of US service in the middle 1960s, but large numbers are in use throughout the word.

Employment

Argentina, Brazil, Greece, Japan, Pakistan, South Africa, Spain, Sweden, Switzerland, Taiwan, Turkey

The 90 mm AA gun M 188

Redeye FIM-43A

United States of America

Missile diameter	70 mm
Missile length	1·22 m
Missile weight	8·2 kg
Guidance system	Infra-red homing
Propulsion	Dual-thrust solid propellant
Maximum effective altitude	2,500 m
Maximum effective slant range	3,400 m
Warhead	HE
Fire control	none

Redeye is designed to provide very low/low level self-defence to forward troops against attacking aircraft. It is robust and simple to operate. Like SA-7 the operator aims the weapon at the attacking aircraft until a buzzer indicates that the infra-red seeking head has locked on to the target's heat source. He then fires it and the missile automatically homes on to the exhaust jets of the aircraft. Redeye was originally intended to be an all-arms weapon but because of problems of fire control it has become a specialist operated weapon as part of the artillery family of air defence weapons.

Employment

Australia	Israel	USA
Denmark	Jordan	
Germany (FRG)	Sweden	

Redeye—missile container

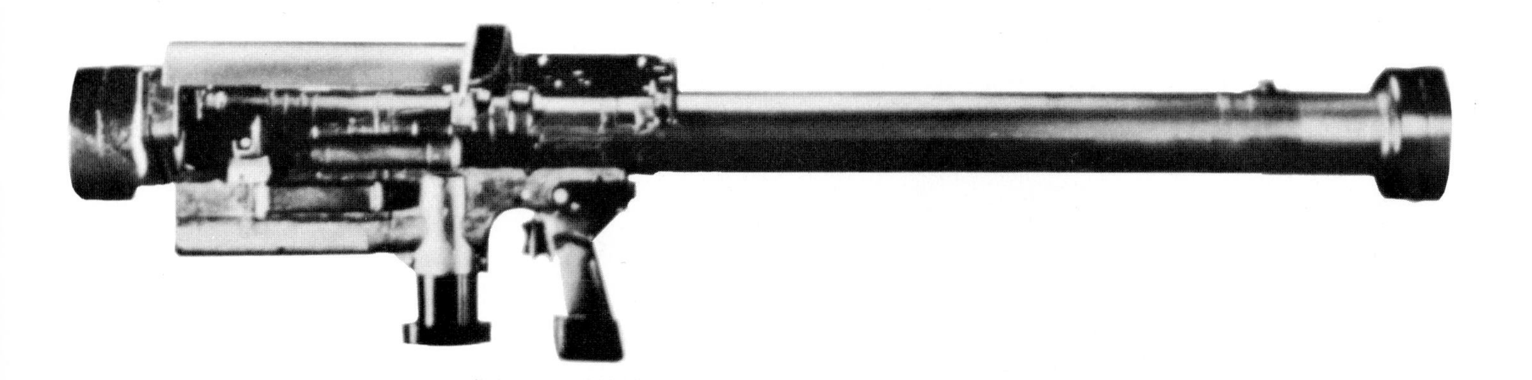

Redeye launcher—rear to the right

SA-7 (NATO code—Grail)

Union of Soviet Socialist Republics

Missile diameter	70 mm
Missile length	1·3 m
Missile weight	9·2 kg
Guidance system	Infra-red homing
Propulsion	Three stage solid propellant
Maximum effective altitude	900 m
Maximum effective slant range	3,000 m
Warhead	HE blast
Fire control	none

The SA-7 is a very low/low level hand held surface to air missile system for self defence against attacking aircraft. It is robust and simple to operate. As with Redeye, the operator only has to aim it at the attacker, until the missile head is receiving a sufficient level of infra-red radiation, and fire it. The missile will then home automatically on to the hot exhaust jets of the enemy aircraft as it recedes. SA-7 first saw active service in Vietnam, and many thousands were fired in the October 1973 War. Its drawbacks are that, being a tail-chaser, it is normally only effective after an attack has been delivered, and it has only a small warhead whose effect can be negated by armour plate over the aircraft's jet pipe.

Employment

Angola	Hungary	Poland
Bulgaria	Iraq	Rumania
Cuba	North Korea	Syria
Czechoslovakia	Kuwait	USSR
Egypt	Libya	Vietnam
Ethiopia	Mozambique	South Yemen
Finland	Peru	Yugoslavia
Germany (GDR)		

SA-7 and operator

Blowpipe

United Kingdom

Missile diameter	76 mm
Missile length	1·4 m
Missile weight	11 kg
Maximum effective altitude	2,000 m
Maximum effective slant range	3,000 m
Warhead	HE with proximity fuse
Fire control	none

Blowpipe was the first man-portable low/very low air defence missile with a true forward-hitting capability. It consists of an aiming unit, which with its IFF facility weighs 7 kg, and a launch canister. The total weight of the system on the operator's shoulder is 21 kg. After an engagement has been completed, the empty canister is discarded, and the aiming unit is clipped to a new missile canister in a few seconds. The aiming unit has a sensor which automatically gathers the missile into the line of sight to the target after launch. The operator then flies the missile on to the target by a joystick control with his thumb. This automatically transmits radio guidance commands to the missile. Blowpipe was originally intended to be an all arms self-defence weapon, but because of the problems of control, it is now operated by artillery air defence batteries. Blowpipe is manned by both regular and territorial artillery air defence batteries. Blowpipe can also be used in an anti-APC role in emergencies.

A version using semi-automatic command to the line-of-sight guidance is being developed, and twin-round towed launchers are being studied.

Employment	Canada UK

Blowpipe missile immediately after launch

Blowpipe—operator awaiting attack

Stinger FIM-92A

United States of America

Missile diameter	70 mm
Missile length	1·52 m
Missile weight	15·1 kg
Guidance system	Infra-red homing
Propulsion	Two stage solid propellant
Maximum effective altitude	4,800 m
Maximum effective slant range	5,000 m
Warhead	HE
Fire control	none

Stinger is being developed as a replacement for Redeye in the very low/low level self-defence role. Although it also uses passive infra-red homing to guide it to the target, it has a much more sensitive homing head than Redeye and it is claimed that this will enable it to detect the heat emitted by an approaching aircraft.

Employment USA

Stinger and operators

RBS 70

Sweden and Switzerland

Missile diameter	106 mm
Missile length	1·32 m
Missile weight	15 kg
Guidance system	Optical laser beam-rider
Propulsion	Two stage solid propellant
Maximum effective altitude	3,000 m
Maximum effective slant range	5,000 m
Warhead	Prefragmented proximity fused
Fire control	PS-70/R radar

RBS has been developed by Sweden in collaboration with Switzerland as a very low/low level self-defence weapon. The sight generates a laser guidance beam which coincides with the line of sight to the attacking aircraft. The missile remains in the centre of the beam after launch until it hits the aircraft—all the operator has to do is to keep the sight accurately on the target. The sight (35 kg) clips on to the missile container, which is then attached to a lightweight tripod stand (22 kg). This provides a seat and a steady aiming position for the operator. Like Blowpipe, the RBS 70 is a true self-defence weapon, in that it can engage approaching aircraft before they have reached the point of weapon release. A further advantage it has over the tail-chasers, like Redeye and SA-7, is that there are no practical counter-measures at present to the laser beam guidance it uses.

Employment

Morocco
Norway
Sweden
United Arab Emirates

RBS 70 showing tripod support

SA-9 (NATO code—Gaskin)

Union of Soviet Socialist Republics

Missile diameter	110 mm
Missile length	1·8 m
Missile weight	30 kg
Guidance system	Infra-red homing
Propulsion	Three stage solid propellant
Maximum effective altitude	5,000 m
Maximum effective slant range	6,000 m
Warhead	HE blast
Fire control	By GunDish radar on ZSU-23-4

SA-9 is a mobile very low/low level missile system, which is being introduced in the Soviet Army to complement the ZSU-23-4 gun for forward area defence. When in position it uses the ZSU-23-4's Gun Dish radar for local warning and fire control. Four or eight missiles are mounted on a modified BRDM-2 with twin elevating launchers either side of the operator's sight. The turret has a 360° traverse. The missile appears to be an improved version of the SA-7 and it probably uses the same infra-red homing head.

Employment Egypt
Syria
USSR (and other Warsaw Pact countries)

SA-9— Gaskin'

SA-8

Union of Soviet Socialist Republics

Missile diameter	210 mm
Missile length	3·20 m
Missile weight	About 200 kg
Guidance system	Probably semi-automatic command to line of sight
Propulsion	Probably two stage solid propellant
Maximum effective altitude	Estimated 6,000 m
Maximum effective slant range	Estimated 12,000 m
Warhead	HE with proximity fuze
Fire control	Radar

SA-8 is a low level air defence system designed to bridge the gap between the very low/low level SA-7 and SA-9 and the low/medium level SA-6. It first appeared in public in the November 1975 parade in Moscow, and it is now in service in the Soviet Army. The entire system, consisting of four missiles, search and tracking radars, and a fire control unit, is mounted on one six-wheeled vehicle. The positioning of the radar dishes and antennae indicates that its guidance system is probably semi-automatic command to line of sight similar to those of Roland and Rapier and its performance is likely to be much the same. The SA-8 appears to have an all-weather 24 hour capability, and it makes a useful addition to the already powerful Soviet air defence inventory.

Employment USSR

SA-8 low level air defence system, on parade

Chaparral MIM-72

United States of America

Missile diameter	127 mm
Missile length	2·91 m
Missile weight	85 kg
Guidance system	Infra-red homing
Propulsion	Solid propellant
Maximum effective altitude	2,000 m
Maximum effective slant range	4,000 m
Warhead	HE
Fire control	Forward Area Alerting Radar system

Chaparral is designed for low/very low level air defence and in the US Army it is used in conjunction with the Vulcan gun system. Four missiles are mounted on a rotating launcher turret on the M-730 self-propelled tracked vehicle. The missile is basically the supersonic air-to-air Sidewinder, modified for ground-to-air use and it is stored and handled an as assembled round of ammunition. It is aimed by the gunner through an optical sight until the infra-red homing head has locked onto the heat source of the target's jet efflux. After launch it automatically homes onto its target. The latest version, the MIM-72C, has an all-accept seeker, allowing it to engage approaching targets.

The M-730 can be made amphibious in a short time by adding a simple swimming kit. It carries a five man detachment and can travel at 40 mph on a road. In the US Army the Forward Area Alerting Radar system provides local warning for Chaparral and Vulcan.

Employment

Israel	Taiwan	USA
Morocco	Tunisia	

Chaparral low level air defence system mounted on XM 730

Tigercat

United Kingdom

Missile diameter	190 mm
Missile length	1·48 m
Missile weight	68 kg
Guidance system	Command to line of sight
Propulsion	Two stage solid propellant
Maximum effective altitude	3,000 m
Maximum effective slant slant range	5,000 m
Warhead	HE
Fire control	Radar

Tigercat is a trailer mounted version of Seacat for battlefield low level air defence. The system consists of a launcher, which has three missiles, and a tracker/director. The operator guides the missile to the target with a thumb stick with radio commands being automatically passed to the missile. Tigercat can be given a 24 hour capability by the addition of the Marconi ST 850 radar which is also mounted on a trailer.

Employment		
	Argentina	Qatar
	India	South Africa
	Iran	UK

Tigercat, test firing

Rapier

United Kingdom

Missile diameter	127 mm
Missile length	2·24 m
Missile weight	42·5 kg
Guidance system	Semi-automatic command to line of sight
Propulsion	Two stage solid propellant
Maximum effective altitude	3,000 m
Maximum effective slant range	7,000 m
Warhead	HE
Fire control	Radar

Rapier air defence system—tracker and operators in foreground

Missile launched from tracked Rapier

Rapier is a lightweight low level air defence system. In the British Army it is mounted on a trailer towed by a 1 tonne landrover and there is a second vehicle in the firing unit towing a trailer containing additional missiles. The Rapier system consists of a launcher, which mounts four missiles two each side of the search and acquisition radar dome, an optical tracker and a small generator. These three components are connected together by cables. The detachment has five men, but once the system is in action the engagement sequence is controlled entirely by one tracker operator. Once the search radar has acquired a target and recognised it as hostile, it automatically alerts the tracker operator and directs the optical tracker until the aircraft can be seen in sight. All the operator then has to do is to keep the tracker's crosswires on the target with a joystick control and to fire a missile, when the computer indicates that the target is within range for effective engagement. The semi-automatic command to line of sight guidance is achieved by a television camera, which tracks the missile's flare, collimated with the optical tracker. The guidance commands to bring the missile on to the line of sight to the target are automatically computed and transmitted to the missile.

A blindfire tracking radar has been developed as an addition to the optical system, and this will provide a full 24 hour capability. It will be mounted on a similar trailer to that of the launcher.

One of the features of the Rapier system is the revolutionary test and repair vehicles, which have been developed to permit maintenance and repair in the forward areas. These use computerised test facilities mounted in a landrover (the Forward Repair and Test Vehicle) and in 4-ton trucks at battery level, one for electronic and electrical repairs and one for optics and hydraulics.

TRACKED RAPIER. A tracked version of the optical system is being developed for the British Army. The tracked Rapier mounts 8 missiles, 4 on each side of the surveillance radar dome.

Employment

Australia	Oman	UK
Brunei	United Arab Emirates	Zambia
Iran		

Roland

France

Missile diameter	160 mm
Missile length	2·40 m
Missile weight	63 kg
Guidance system	Semi-automatic command to line of sight
Propulsion	Two stage solid propellant
Maximum effective altitude	3,000 m
Maximum effective slant range	6,200 m
Warhead	HE with proximity fuze
Fire control	Radar

Roland is an antonomous low-level air defence system. It will be mounted on the Marder or AMX 30 chassis in the West German and French Armies, and the US Army mounts it on the M 109 tracked vehicle. The pulse doppler surveillance radar detects and identifies the target for the operator to fire and track the missile through his optical sight. The missile is automatically guided on to the line of sight to the target by an infra-red guidance system similar to that used for the HOT anti-tank guided weapon. Roland I will be introduced first, as a daylight only system. Roland II will be a 24 hours blindfire system, with a radar tracker being used in the place of the optical sight.

Employment

Brazil	Germany (FRG)	USA
France	Norway	

Roland low level air defence system

Crotale

France

Missile diameter	156 mm
Missile length	2.93 m
Missile weight	85·1 kg
Guidance system	Semi-automatic command to line-of-sight
Propulsion	Single stage solid propellant
Maximum effective altitude	3,600 m
Maximum effective slant range	8,500 m
Warhead	HE with proximity fuze
Fire control	Radar

Crotale has been developed by France as a low-level air defence system. The French Army has preferred Roland for this role, but South Africa has a version of Crotale in service called Cactus. Four missiles and a command guidance radar are mounted on a rotating launcher on a lightly armoured wheeled vehicle. A similar vehicle mounts a Thompson pulse-doppler S band surveillance radar, which has an automatic target evaluation system. Data is passed from the surveillance radar to the launcher by cable for distances up to 400 m, or by radio up to 5,000 m. A version mounted on AMX 30 chassis has been developed for Saudi Arabia, where it is known as Shahine.

Employment

Egypt	Libya	South Africa
France	Morocco	Spain
Greece	Pakistan	United Arab Emirates
Kuwait	Saudi Arabia	

Crotale low level air defence system

Indigo-MEI

Italy

Missile diameter	195 mm
Missile length	3·3 m
Missile weight	120 kg
Guidance system	Radio command/beam-riding
Propulsion	Single stage solid propellant
Maximum effective altitude	5,000 m
Maximum slant range	10,000 m
Warhead	HE fragmentation with proximity fuse
Fire control	Radar

Italy is developing Indigo as a vehicle-mounted low-level air defence system. The system consists of 4 vehicles based on modified M 548 chassis. One carries search and tracking radar, two are fitted with 6 ready-to-fire missiles, and the fourth transport 12 reload rounds. The primary method of guidance is beam-riding on the line of sight to the target, but radio command guidance with infra-red tracking is provided as a back up, if for some reason the beam-riding fails or is effectively countered.

Employment Development in Italy

Spada

Italy

Missile diameter	203 mm
Missile length	3·7 m
Missile weight	220 kg
Guidance system	Semi-active radar homing
Propulsion	Single stage solid propellant
Maximum effective altitude	—
Maximum effective slant range	18 km approx
Warhead	HE with proximity fuze
Fire control	Radar

The Spada air-defence system uses the Selenia Aspide missile which is additionally employed as an air-to-air weapon and in the Albatros naval system. Spada is planned to defend Italian Air Force bases from aircraft attacking at medium and low altitudes. Up to four fire sections, each with up to three quadruple launchers, will be assigned to a central detection centre.

Employment Italy (under development)

SA-3 (NATO code—Goa)

Union of Soviet Socialist Republics

Missile diameter	450 mm
Missile length	6·7 m
Missile weight	950 kg
Guidance system	Radar command guidance
Propulsion	Two stage solid propellant
Maximum effective altitude	15,000 m (estimated)
Maximum effective slant range	35,000 m (estimated)
Warhead	60 kg HE blast
Fire control	Low Blow

The SA-3 is a low/medium level system with a pair of missiles carried on a six-wheeled ZIL 157 truck and fired from a two-round launcher. Fire control is provided by the NATO code Low Blow radar in association with a further surveillance radar. It provides low/medium level cover at divisional level and is the link between the SA-2 and SA-4 and the forward area SA-6. SA-3 has been in service since about 1960.

Employment

Afghanistan	Egypt	Poland
Albania	Germany (GDR)	Rumania
Algeria	Hungary	Syria
Bulgaria	India	USSR
China	Iraq	Vietnam
Cuba	North Korea	Yugoslavia
Czechoslovakia	Libya	

Two missiles on launcher

Hawk/Improved Hawk MIM 23A/MIM 23B

United States of America

Missile diameter	350 mm
Missile length	5·12 m
Missile weight	580 kg
Guidance system	Semi-active homing
Propulsion	Two stage solid propellant
Maximum effective altitude	15,000 m (estimated)
Maximum effective slant range	20,000 m (estimated)
Warhead	HE blast fragmentation
Fire control	Pulse and continuous wave acquisition radar

Hawk (Homing All the Way Killer) is a highly mobile low/medium level air defence system, which can engage supersonic aircraft up to an altitude of 18,000 m. Three missiles are mounted either on a two-wheeled trailer or on the self-propelled tracked M-727. Ground support in a Hawk battery consists of a pulse acquisition radar, a continuous wave acquisition radar, a range-only radar, two illuminating radars, six three-missile launchers, a tracked missile loader-transporter and the command post. When the acquisition radars have detected a target, it is illuminated by one of the continuous wave illuminating radars. After launch the missile tracks the electromagnetic energy which is reflected from the illuminated target.

A later version known as Improved Hawk has a new guidance package, a larger warhead and a better propellant. Solid state components reduce the maintenance load in the field.

Employment

Belgium	Italy	Philippines
Denmark	Kuwait	Saudi Arabia
France	Japan	Spain
Germany (FRG)	Jordan	Sweden
Greece	South Korea	Taiwan
Iran	Laos	Thailand
Israel	Netherlands	USA

Hawk mounted on XM 727.

SA-2 (NATO code—Guideline)

Union of Soviet Socialist Republics

Missile diameter	500 mm	
Missile length	10·6 m	
Missile weight	2,300 kg	
Guidance system	Radio command	
Propulsion	Liquid propellant with solid fuel boosters	
Maximum effective altitude	28,000 m	(estimated)
Maximum effective slant range	50,000 m	(estimated)
Warhead	130 kg HE blast	
Fire control	Song radar	

SA-2 is a medium/high level surface-to-air missile mounted on an articulated trailer, towed by a ZIL 157 truck. It has been in service since the 1950's and there are already several versions of it, possibly with terminal guidance, in the Soviet Army. It has poor cross-country ability and would not be expected to be deployed in forward areas.

Employment

Afghanistan	Egypt	Mongolia
Albania	Germany (GDR)	Poland
Algeria	Hungary	Rumania
Bulgaria	India	Syria
China	Iraq	USSR
Cuba	North Korea	Vietnam
Czechoslovakia	Libya	Yugoslavia

The missile on its launcher

SA-2—Guideline—on its transporter

SA-6 (NATO code—Gainful)

Union of Soviet Socialist Republics

Missile diameter	335 mm
Missile length	6·2 m
Missile weight	550 kg
Guidance system	Command with semi-active radar homing
Propulsion	Ramjet with solid fuel booster
Maximum effective altitude	15,000 m (estimated)
Maximum effective slant range	35,000 m (estimated)
Warhead	80 kg HE blast with proximity fuze
Fire control	Straight Flush radar system

The SA-6 is a low/medium level surface-to-air missile mounted on a tracked launcher vehicle. It is used to provide comprehensive air defence cover of forward areas, together with the very low/low level SA-7 and 9, and the S-60 and ZSU-23-4 gun systems. Fire control is provided by the Straight Flush radar and a separate surveillance radar and height finder. The SA-6 was the most successful air defence missile in the October 1973 war. It is equivalent to the US Hawk.

Employment

Bulgaria	Iraq	Rumania
Czechoslovakia	Libya	Syria
Egypt	Mozambique	USSR
Germany (GDR)	Poland	Vietnam
Hungary		

Foreground, Straight Flush radar—SA 6 in background

SA-4 (NATO code—Ganef)

Union of Soviet Socialist Republics

Missile diameter	800 mm	
Missile length	9 m	
Missile weight	2,000 kg	
Guidance system	Radar command guidance	
Propulsion	Ramjet with 4 solid fuel boosters	
Maximum effective altitude	30,000 m	(estimated)
Maximum effective slant range	70,000 m	(estimated)
Warhead	HE blast	
Fire control	Pat Hand	

The SA-4 is a medium/high level surface-to-air missile system with two missiles mounted in tandem on a heavy tracked launcher vehicle. It is easily distinguishable by its four wrap-round boosters, which are similar in appearance to those on Bloodhound. Its C-band Pat Hand radar provides fire control in association with surveillance and height-finding radars. All its radars are mounted on heavy tracked vehicles. Despite its impressive size SA-4 appears to have good cross-country mobility and could on occasions be deployed at brigade/regimental level. It is approximately equivalent to US Nike Hercules.

Employment Czechoslovakia, Germany (GDR), USSR

SA-4—Ganef

The tracked launcher in a hide being loaded with two SA-4 missiles

Bloodhound II

United Kingdom

Missile	550 m
Missile length	7·75 m
Missile weight	2,454 kg
Guidance system	Semi-active homing
Propulsion	Ramjet with four solid propellant boosters
Maximum effective altitude	Estimated 30,000 m
Maximum effective slant range	Estimated 80,000 m
Warhead	HE with proximity fuze
Fire control	Target illuminating radar

Bloodhound II replaced the earlier version in the mid 1960s as a medium/high level air defence system. The semi-active homing system requires the target aircraft to be illuminated by an X band continuous wave-doppler radar during the missile's flight. It is primarily a static weapon system for asset defence with the long range Type 87 target illuminating radar, but there is also a trailer mounted version with a mobile Type 86 Indigo Corkscrew radar, which is the same as the British Army Radar No 10 used with Thunderbird II. Each launcher has one missile. One target illuminating radar, with its associated computers in a command post, can control up to four missiles.

Employment	Singapore	Switzerland
	Sweden	UK

Bloodhound II

Nike Hercules

United States of America

Missile diameter	880 mm
Missile length	12·7 m
Missile weight	4,800 kg
Guidance system	Command
Propulsion	Two stage solid propellant
Maximum effective altitude	35,000 m (estimated)
Maximum effective slant range	150,000 m (estimated)
Warhead	HE or nuclear
Fire control	Radar

Nike Hercules is the principal NATO medium/high level air defence system. Like its predecessor Nike Ajax, many Nike Hercules are in fixed sites defending cities and military installations, but those which provide cover for the field army have road mobility. A Nike Hercules battery's ground equipment consists of a high and a low power acquisition radar, a target tracking radar, a missile tracking radar, data processing computers, a number of remotely controlled launchers and electronic counter measures equipment. The target is first detected by the acquisition radar and then pinpointed for interception by the target tracker. The missile is guided to the target by the interplay of data from the target and missile tracking radars. Nike Hercules was first introduced at the end of the 1950s and Patriot is being developed as its successor.

Employment

Belgium	Italy	Norway
Denmark	Japan	Taiwan
Germany (FRG)	South Korea	Turkey
Greece	Netherlands	USA

Nike Hercules: left foreground

Patriot

United States of America

Missile diameter	410 mm
Missile length	5·2 m
Missile weight	1,000 kg
Guidance system	Command with semi-active homing
Propulsion	Single stage solid propellant
Maximum effective altitude	24,000 m (estimated)
Maximum effective slant range	60,000 m (estimated)
Warhead	HE or nuclear
Fire control	Radar

Patriot is being developed as a medium/high level air defence system to replace Nike Hercules and Hawk. The launcher and its ground-support equipment will be mounted on tracked vehicles and so it will be much more mobile than Nike. The prototype has six missiles on a launcher and a possible organisation would provide eight launchers per battery. One phased-array radar is used for surveillance, target acquisition and target tracking. This one radar performs the tasks which require five radars in a Nike battery and four radars in a Hawk battery.

Miniaturization and advances in the art of solid-state components have made possible the extremely compact organization of a Patriot battery with only a dozen or so vehicles involved in the firing sequence, and far better reliability in the field.

Employment Development in USA

Patriot (formerly Sam- D) - a test firing

X. Air Defence Ancillary Equipment

All but the smallest air defence weapons operate at a range which is too great for accurate visual control. Furthermore, the weapon must be able to react to the modern aircraft's all-weather day and night capability and its speed which brings it into range too fast for visual target acquisition.

A number of tactical radars are available with both a search and tracking capability. In this section, only radars which are used for gun and smaller missile control are included. Those which are an integral part of the larger missile systems, have already been described with the missile concerned.

Eldorado (TH-D-1229)

France

Eldorado is a one man operated, trailer-mounted air defence fire control system. The trailer, with roof-mounted radar antenna and its own power pack, is self-sufficient, but weighs only 1,600 kg. An optical system works in parallel with the radar, and both systems are displayed simultaneously to the operator. The equipment has an inbuilt target acquisition capability, but is designed to operate with Mirador, a specialist acquisition radar, whose circuitry is built into Eldorado, and whose antenna and tripod fit into the Eldorado trailer.

Eldorado trailer

Mirador II

France

General. This S-band coherent pulse radar is primarily a target acquisition system for use with guns or missiles. It has high resistance to electronic counter-measures. It can be used in a tracking mode while scanning.
Capabilities. Mirador II is effective against low and very low targets. It produces a 360° search pattern, the radar head rotating at one revolution per second. Maximum detection range is 18 km. It resolves range to 400 m, or 200 m when tracking, and is accurate to 0·75° in azimuth.
Operation. An audible alarm system indicates the presence of a target. Target data can be passed to several fire positions. A plan position indicator presents the operator with radial speed and range information to enable a speedy threat evaluation.

Mirador aerial and tripod

L 4/5 weapon control system

Netherlands

General. The L 4/5 is a system which dates from the early 1960's and is designed for gun or missile control. The complete system is mounted on a four-wheel trailer from which a canvas extension is deployed in action. There are three antennae, two for search and one for tracking. IFF is fitted, together with numerous anti-jamming features. The complete system weighs 6,000 kg, and is operated by two or three men.
Capabilities. Each L 4/5 can control up to three guns or missiles simultaneously. The two search antennae are mounted back to back and search the medium/low, and very low height bands respectively. The search antennae can continue to operate while the tracking radar is in use. A back-up optical sight is fitted. Maximum range is 30 km, and the system can handle target speeds up to 500 m/sec.
Operation. The rapid 360° spiral search pattern of the search radar produces a continuous plan position indicator display in the control cabin. Sector or continuous search can be selected. Targets are picked up for tracking automatically, or manually.

The L 4/5 system deployed for action

VL 4/41 fire control system—'Flycatcher'

Netherlands

General. Flycatcher is a new X-band radar equipment which is adapted from the fire control system fitted to the Gepard. It can control up to three guns, or two guns and one missile. It is housed in a container with the antennae normally carried internally but mounted on the roof in action. A television camera can be mounted with the tracking antenna.
Capabilities. The range is 30 km. Target speed limitations are 500 m/sec horizontally, and 300 m/sec vertically. The computer produces firing data, which includes ballistic corrections for the gun's performance.
Operation. The appearance of a target in the search pattern triggers an alarm. Targets are automatically interrogated by IFF.

Flycatcher, showing entrance to cabin and generator

Ericsson/Arenco radar system

Sweden

General. L. M. Ericsson have produced several X-band air defence fire control radars. The PE-48/T is used in conjunction with an Arenco computer.

Capabilities. The system is a tracking radar which produces firing data for two 40 mm or 57 mm guns. It can also be used in the search mode.

PE-452/T radar system

Sweden

General. This is an elderly radar/optical system. The PE-452/T works in conjunction with a Contraves optical tracker: the X-band radar measures range, while the optical system produces angular measurements. A computer combines both inputs into gun firing data.

Capabilities. Being partly optical, this system is not suited to all-weather day and night operation. Two 40 mm guns can be controlled.

Skyguard fire control system

Switzerland

General. Skyguard can be used to control both guns and missiles. The system consists of a search radar, tracking radar, and television tracking. A Contraves Core II M digital computer is used. Both radars are of the pulse-doppler type, and operate independently. A generator is built into the trailer which contains the system.

Capabilities. The pulse-doppler search radar features high clutter suppression, high resistance to electronic counter-measures, and automatic target transfer to the tracking radar. A target alarm is fitted. The computer is used to assess the threat, produce target data for the guns and missiles, test the operation of the equipment and produce electronic training signals. There is provision for IFF. Range is 20 km and range accuracy is 160 m.

Operation. The plan position indicator can be used to present target displays from 0·3 to 20 km in range. One or two operators are needed depending on the tactical situation.

Skyguard low-level radar and fire-control system

Superfledermaus fire control system

Switzerland

General. Superfledermaus is designed to control three missiles or three guns of 35–40 mm calibre. It has X-band acquisition and tracking radars, and a back-up telescopic sight for the latter. It is mounted on a four-wheel trailer, weighs 5,000 kg, and has been sold to over twenty countries. Data production is computerised and a constant muzzle velocity measurement facility at each gun is included in the fire control system.
Capabilities. Using a helical or sector scan, targets can be acquired up to 50 km. Autotrack is effective to 40 km. The computerised data production incorporates correction for gun parallax, wind speed and air density as well as muzzle velocity.
Operation. There is autotransfer from acquisition to tracking radars with a plan position indicator and range/height indicator to present the maximum tactical information to the operator.

Superfledermaus: side view showing detail of the trailer

'Fire Can' fire control radar

Union of Soviet Socialist Republics

General. Fire Can is an elderly S-band radar with a search and tracking capability. It is trailer mounted, and has a single 1,500 mm dish antenna. The dish is perforated to reduce weight.
Capabilities. Fire Can can acquire targets to a range of 80 km. The maximum tracking range is 35 km and accuracy in the order of 15 m is probable.
Operation. In the search mode all targets are indicated on a plan position indicator. Target selection, and transfer to the tracking mode is manual. This radar is used to control 57 mm and 85 mm guns.

XI. Coast Artillery

Coast artillery as a specialised branch fell into disuse after World War II. Fortresses with fixed armaments from 6 inch to 15 inch in concrete emplacements designed to protect naval bases and anchorages proved only too vulnerable to land attack and aerial bombing. Equally the supporting fire of naval craft and aerial bombardment effectively neutralised coastal fortifications sited to cover beaches and landing places suitable for invasion. While it was realised that coast artillery was to some extent a deterrent it locked up men and equipment in fixed defences which in the event seldom if ever fired a shot. Further, the advance of weapon technology in the shape of airpower and surface-to-surface guided missiles specially designed for attacking ships provided a far more flexible and effective counter both to invasion and to long-range ship-to-shore bombardment.

It is always difficult to draw boundaries between weapon systems but the anti-ship missiles are essentially automatically piloted aircraft designed to fly at altitudes as low as two metres above the sea level and are not, as far as is known, normally handled by artillery personnel although all are potentially shore-to-ship weapons and two, the Matra and Oto Melara OTOMAT (France and Italy) and the Swedish Saab RB-08, can be used in this role. Similarly any naval gun system can be adapted for shore-to-ship role without major modification. None of these weapons is included in this chapter.

Nevertheless guns, especially if sited in commanding positions, are still an economical method of providing close-in defence of harbour and anchorages against surprise attack or fast, small raiding craft and also for the control of straits and approaches through shoals or shallows where shipping must follow a predetermined course. 'Classical' coast systems were very effective. The range was discovered by continuously solving the triangle formed by the height of the gun above sea-level and the angle of sight from the gun to the target and range and bearing data was passed electronically to the split function layers by the 'follow-the-pointer' system. Even with manually operated breech mechanisms laying direct and estimating the lead, the chance of a hit in two or three rounds was good. Some of these antiquated but still servicable systems of fire control may still be in service for harbour defence. Their disadvantage is that they are all optical and depend on good visibility, and the director operators and layers have to be skilled, intelligent and regularly practised.

Anti-aircraft guns, if they can be modified to shoot with angle of sights of depression with their radars and fire-control instruments are effective all-weather coast defence artillery. The German 88 mm Flak 36 is used in this role in Yugoslavia. Reportedly such coast batteries as are known are adaptations from other roles. Only the Swedish firm of Bofors has designed guns specifically for coast defence; the 75 mm described here and a 120 mm.

According to *The Military Balance* (published annually by the International Institute for Strategic Studies, London) The Soviet Union, Norway, Sweden, Yugoslavia and Cuba have coast artillery units which are naval manned. The most complete, as might be expected, are the Soviet Union's, covering all naval bases and major ports as well as providing full visual and radar cover along the coast line. Cuba is equipped with the Samlet missile. In Finland, Portugal and Spain the coast artillery is manned by the army. Albania and Egypt may also maintain some coast defence units.

75 mm coast defence gun L/60

Sweden

Calibre	75 mm
Barrel length	60 calibres
Muzzle brake	Nil
Weight	—
Ammunition	HE (5·52 kg)
	AP (5·73 kg)
Rate of fire	25 rds/min
Muzzle velocity	870 m/sec
Maximum range	12,000 m
Elevation limits	−5° +20°
Traverse on carriage	360°
Detachment	5

The Swedish coast defence system is controlled by the Navy. It consists of 20 batteries equipped with guns, and RB 08 and SS 11 missiles. This gun is made by Bofors and mounted in fixed emplacements which have an armoured dome type turret built over a bunker which houses ammunition.

The monobloc barrel is fitted with a fume extractor but no muzzle brake. The vertically sliding breech block opens downwards, a design feature which helps to reduce the height of the turret to a minimum. The barrel is mounted into a 'cradle' structure which extends rearwards and houses the automatic loading device and recoil system, and also counter balances the elevating mass. Thus only the barrel projects from the turret.

Loading is fully automatic, operated partly by recoil action, and partly by electric power. The fixed ammunition is stored in a magazine chamber beneath the turret.

An electrically operated rod-type hoist moves the cartridges up one step for each movement of the hoist driving mechanism which consists of a worm and bevel gear driven via a connecting rod linkage. The movement of the hoist is controlled by cam-actuated microswitches, which control the entire functioning of the hoist. In an emergency the hoist can be operated manually.

The cartridges, which are transported in the hoist in an upright position, are transferred to the loading channel by means of a loading tube. During the hoisting procedure, the loading tube is hooked up vertically over the hoist drum, and when a cartridge has been hoisted up into the loading tube, the tube is automatically

Turret and cupola—L/60 system. Note fume-extractor on barrel

released and manually turned in to the loading channel in front of the rammer. The loading tube is supported on a pendulum which moves around the elevation trunnion centre. Raising the pendulum and loading tube with cartridge is effected by means of a torsion spring which is cocked by the recoil movement. The gun has a spring-operated rammer, which is also cocked by the recoil. The rammer is released automatically when the loading tube is swung up to the loading channel.

Split function, manual laying is used. For indirect laying, 'follow-the-pointers' are used. For direct aiming there is a periscopic sight with dual eyepieces, one for the elevation layer and one for the traversing layer. The tangent elevation is set manually by means of a handwheel, which actuates scales in the focus of the object lens of the periscopic sight. The gun is fired mechanically by means of a pedal located at the traversing layer's position.

The entire installation weighs 17,500 kg, which includes the 5,250 kg turret dome with its 70 mm of armour. The rate of fire is very high, but is limited by the speed of operation of the ammunition hoist: when operated manually only 15 of the 11·5 kg rounds can be lifted per minute.

Employment Sweden

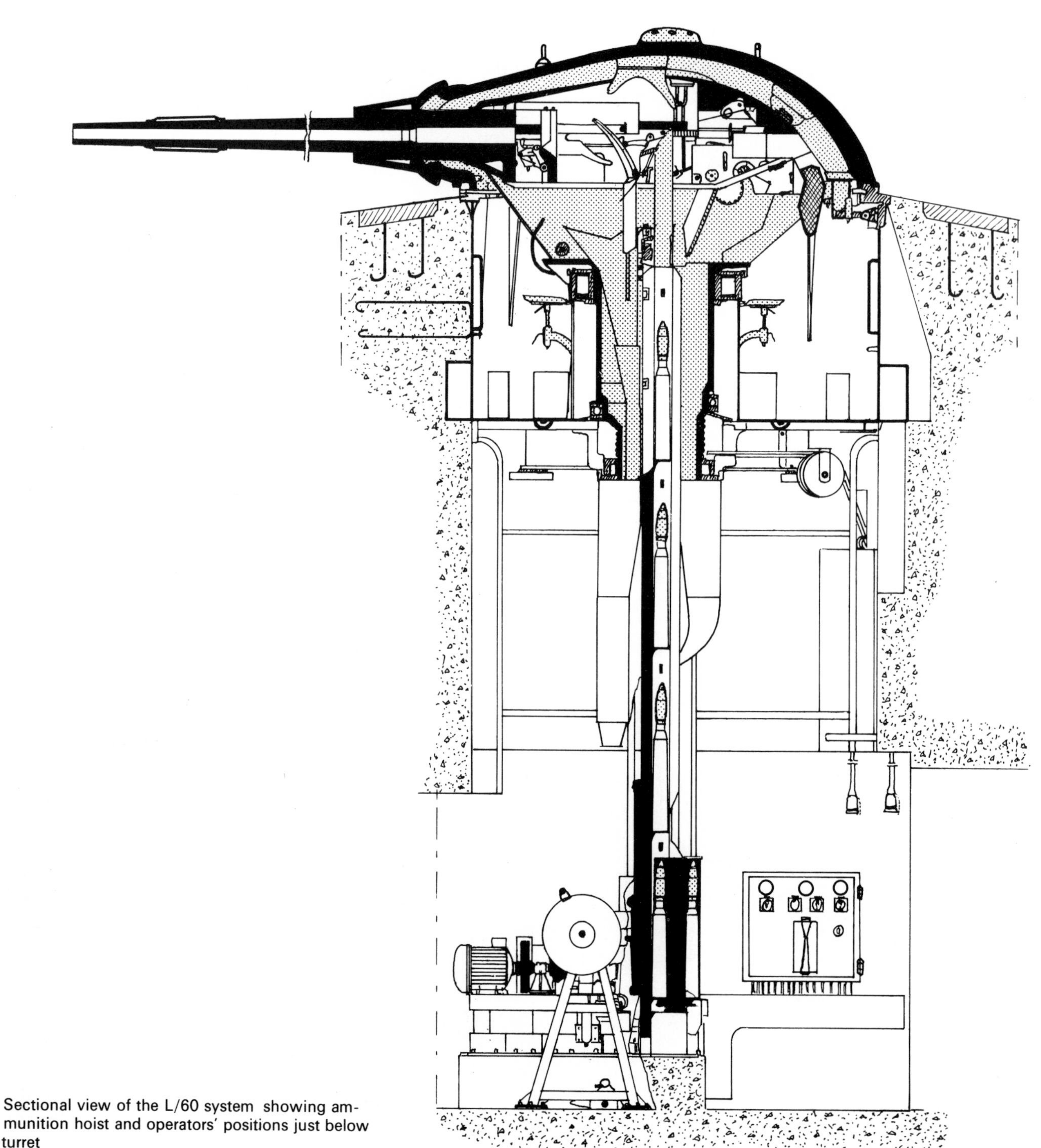

Sectional view of the L/60 system showing ammunition hoist and operators' positions just below turret

88 mm coast defence gun Flak 36

Germany

Calibre	88 mm
Barrel length	56 calibres
Muzzle brake	Nil
Weight	5,100 kg
Ammunition	HE (8·9 kg) AP HVAP
Rate of fire	15–20 rds/min
Muzzle velocity	820 m/sec
Maximum range	14,860 m
Elevation limits	—
Traverse on carriage	—
Detachment	—

This is the famous Krupp anti-aircraft gun of World War II, now relegated to a coast defence role, and that only by Yugoslavia. It is fitted with a semi-automatic loading device which produces the high rate of fire. This feature, together with the 'follow-the-pointers' sights, is retained from the original anti-aircraft mounting and makes the gun especially effective against moving targets.

The guns are mounted in permanent fixed emplacements along the Adriatic coast.

Employment Yugoslavia

130 mm coast defence gun

Union of Soviet Socialist Republics

Calibre	130 mm
Barrel length	55 calibres
Muzzle brake	Pepperpot
Weight	19,000 kg
Ammunition	HE (33·4 kg) APHE (33·6 kg)
Rate of fire	5–10 rds/min
Muzzle velocity	930 m/sec
Maximum range	27,000 m
Elevation limits	−5° +45°
Traverse on carriage	360°
Height on wheels	2·8 m
Detachment	—

This gun is thought to be the M 1936 naval gun mounted on a World War II German carriage. The gun's ancestry, muzzle velocity and shell weight are common to the M 1946 field gun, but the range quoted differs. It is also related to the 130 mm anti-aircraft gun M 55.

The elderly carriage of this coast defence version consists of a turntable platform mounted on four double road wheels; the wheels, axles, and tow bar are removed in action; these parts alone weigh 3,000 kg. The barrel, with the recoil mechanism beneath it, is mounted high on the carriage in a massive cradle. A large curved shield is fitted.

The quoted rate of fire is high for the size of the gun and suggests that loading may be automated or mechanically assisted. The loading system used for the 130 mm anti-aircraft gun would seem suitable.

Employment Egypt
USSR

130 mm coast defence gun in its travelling position

Obsolete Equipments

Many countries, especially those which cannot afford the high costs of modern equipments, retain either in reserve stock or in service guns which, although they may be obsolete, are nonetheless lethal. They are useful for training and cheaper for practising observation of fire, since old stocks of ammunition can be used up. It is difficult to be positive about the status of these weapons, but the table below shows equipments which may be found on inventory or in service.

Note: *Field Guns and Howitzers.* Where there are amunition options, the *Muzzle velocity* and *Ammunition* refer to HE.

Anti-tank Guns. Where there are ammunition options, the *Muzzle velocity* and *Ammunition* refer to HVAP. *Range* is Maximum Effective Range.

Anti-aircraft Guns. Rate of fire is Cyclic Rate of Fire: *Range* is Maximum Slant Range.

	Calibre (mm)	Rate of fire (rds/min)	Weight (kg)	Range (m)	Muzzle velocity (m/sec)	Ammunition (kg)	Traverse	Elevation	Country of origin
Field Guns and Howitzers									
Field gun M 1897	75	12	1,160	11,200	550	7·24	6°	−11° + 18°	France
Mountain gun Type 94	75	10–12	535	8,300	385	6·3	45°	−10° + 45°	Japan
Field gun 75/22	75	—	764	10,500	480	—	7°	0° + 40°	Spain
Howitzer M 1902	75	—	—	—	—	—	—	—	Sweden
Pack howitzer M 116	75	3	653	8,280	384	6·62	6°	−5° + 45°	USA
Divisional gun M 1942 (ZIS 3)	76	20	1,116	13,300	680	6·2	54°	−5° + 37°	USSR
Gun M 1092/30	76	8	1,350	12,000	717	6·4	2°	−5° + 37°	USSR
Howitzer M 1927	76	14	780	8,570	387	—	6°	−6° + 25°	USSR
M 36	76	—	1,620	13,600	706	6·4	60°	−5° + 75°	USSR
Mountain gun M 1938	76	14	786	10,000	503	6·23	10°	−8° + 70°	USSR
Howitzer M 1943	76	14	600	4,200	—	—	60°	−8° + 25°	USSR
25 pdr gun/howitzer	88	5	1,741	12,250	520	11·34	360°	−5° + 45°	UK
M 37–10	105	—	—	—	—	—	—	—	USSR
Field gun M 1941	105	7	1,800	13,400	600	14·9	53°	−6° + 45°	USSR
Light field howitzer M 61/37	105	7	1,800	13,400	600	14·9	53°	−6° + 45°	Finland
Light field howitzer M 18/40	105	6–8	1,800	12,325	540	14·8	60°	−5° + 45°	Germany

	Calibre (mm)	Rate of fire (rds/min)	Weight (kg)	Range (m)	Muzzle velocity (m/sec)	Ammunition (kg)	Traverse	Elevation	Country of origin
Light field howitzer M/40	105	10	1,850	14,300	620	14·5	50°	−5° + 45°	Sweden
Field gun M 1937 (A 19)	122	5-6	7,117	20,800	800	25·5	58°	−2° + 65°	USSR
5·5 in gun	140	3	5,850	16,400	510	36·3	60°	−5° + 45°	UK
Field howitzer M 39	150	4–6	5,720	14,600	580	41·5	45°	−5° + 66°	Sweden
Field howitzer FH 18	150	5	6,500	15,000	580	42	45°	−5° + 65°	Sweden
Medium field howitzer M 40	150	4	5,530	13,300	520	43·5	60°	0° + 45°	Germany
Howitzer M 18/46	152	4	5,512	12,400	508	40	60°	0° + 45°	Czechoslovakia
Howitzer M 1937 (ML 20)	152	4	7,930	17,300	655	43·6	58°	−2° + 65°	USSR
Howitzer M 1938 (M 10)	152	4	4,150	12,400	508	39·9	50°	−1° + 68°	USSR
Gun M 2	155	1	13,880	23,500	835	57·6	60°	−2° + 63°	USA
8 in howitzer M 115	203	0·5	13,471	16,800	594	90·7	60°	−2° + 65°	USA
Mortars									
M 43	160	3	1,270	5,150	—	40·8	25°	+45° + 80°	USSR
M 53	160	2–3	1,300	8,040	—	40·8	24°	+50° + 80°	USSR
Anti-tank Guns									
Gun M 1942	45	25	570	500	1,070	0·9	60°	−8° + 25°	USSR
Gun M 1943 (ZIS 2)	57	20–25	1,150	500	1,270	1·8	56°	−5° + 25°	USSR
Auxilliary propelled gun M 1955	57	15–20	1,250	500	1,270	1·8	56°	−5° + 15°	USSR
17 pdr gun	76·2	20	3,040	1,200	—	—	60°	−6° + 16·5°	UK
Recoilless rifle M 59A	82	6	386	1,000	745	6	360°	−13° + 25°	Czechoslovakia
Recoilless rifle T 21	82	5–6	20	300	250	2·1	—	——	Czechoslovakia
Gun M 1945 (D 44)	85	20	1,725	1,600	1,030	5	54°	−7° + 42°	USSR
MOBAT	120	—	764	800	462	12·84	360°	−7° + 30°	UK
WOMBAT	120	4	294	1,000	462	12·84	360°	−8° + 17°	UK
SP assault gun ASU 57	57	15–20	5,550	500	1,270	1·8	12°	−5° + 15°	USSR
Anti-aircraft Guns									
Automatic cannon HO 831-SLM	30	600–650	—	3,000	1,080	0·36		— —	Switzerland
3·7 in gun Mk 3A	94	20	8,392	12,000	792	12·6		— —	UK
Heavy gun M 49 (KS 19)	100	15–20	11,000	7,000	900	15·7		— —	USSR
Heavy gun M 55 (KS 30)	130	12	30,000	10,000	950	34		— —	USSR

Glossary of artillery terms

This is intended as a guide to the non-specialist reader. The precision of many once exact terms has been eroded by popular usage; e.g. 'barrage' and 'shrapnel', while others have dropped out of use in an era in which elaborate fire-plans prepared in advance for attack or defence have become unfashionable; e.g. the true 'barrage' and 'counter-preparation'. American and English usages differ and may cause confusion; e.g. what the US artillery calls 'time fire' the British call 'air burst' (i.e. burst in the air with a time or VT fuze) and the British 'time, or timed, programme' is a schedule of fire coordinating the engagement of targets with the projected operations of other troops, usually in the attack, although there are, especially in the Soviet Army, elaborate defensive fire plans. Artillery tactics and design are technical subjects and naturally a technical jargon has grown up to describe them. Only those terms which will assist the ordinary reader to understand the art and in recognition of equipments have been included here.

Accuracy (of fire) The accuracy of a weapon system is the measure of the deviation of the point of aim, expressed in terms of the distance between the point of aim and the mean point of bursts (or MPI).

Adjust Adjusting fire. See *Ranging*.

Aiming Circle (US) See *Director*.

Aiming Point, Aiming Post Natural reference point from which angles in azimuth of lines of fire are measured by dial sight. Aiming posts carried as part of gun stores are stuck in ground as an alternative method.

Ammunition See under special headings: *Round* of, *Fixed, Semi-fixed, Separate, Shot, Shell*, etc.

Angle of Sight (US *Site*) Angle line joining gun to target makes with horizontal.

Angular Measure Sights graduated in the familiar degrees and minutes are now to be found only on obsolescent British equipments such as the 25 pdr. The artillery measure is the *Mil*, of which there are, confusingly, two sorts. The British and the Americans count 6,400 mils as equivalent to 360°. As well there is the 'Continental' mil which is used by the USSR and the WTO artilleries counting 6,000 mils to 360°. Mils are used not only for sight data but compass bearings and all survey and other calculations.

Armour Piercing (AP) High explosive or kinetic energy solid shot. See under *HEAT, HESH, Discarding Sabot*.

Assault Gun Hybrid weapon, tracked and armoured, and used as indirect fire weapons, anti-tank weapons or direct fire artillery weapons. They fall into the class of armoured fighting vehicle (AFV) as well as 'artillery'.

Auxiliary Propulsion Unit Fitted on carriage to some tractor drawn equipments to facilitate movement without having to call up bulky tractor.

Balancing Springs Used to restore balance of barrel and recoil gear assembly when trunnions cannot be placed at the centre of gravity.

Ballistics Science of; concerned with events in bore of gun and also trajectory of *projectile* and its behaviour in flight.

Ballistic—for projectile in free flight as opposed to guided.

Barrage Colloquially synonym for 'bombardment'. Originally standing belt of fire (see *Defensive Fire*) to halt attack. Now moving or rolling belt of fire used to cover advance or assault.

Barrel Of gun, also '*piece*' (trad., but in British use) and '*tube*' (US).

Base Ejection Of 'carrier' shells used, e.g. for delivering smoke canisters, in which a small charge blows off the base and ejects the filling.

Base Plate Steel disc used to support breech end of mortar.

Battalion Artillery unit formed by grouping batteries together (but in British usage this is a 'regiment').

Battery Group of 4 or 6 guns with command post, signal communications etc. calibrated to shoot as one weapon.

Bearing Picket Post or picket marking position from which survey can be initiated, from bearings to natural reference points and grid location provided.

Bomb Projectile fired from mortar, from its shape, resembling aerial bomb.

Bomblet Sub-munitions carried in warhead of shell or missile to obtain scattered effect (see *minelet*).

Breech-Loading Also in British usage synonym for self-sealing.

Breech-Mechanism If breech loading (BL), made up of cylindrical block with interrupted screw-threads to engage breech, gears or linkage and lever to open and shut it, lock containing firing mechanism, and means of obturation, or sealing breech to prevent escape of propellant gases. If quick-firing (QF) or cartridge-sealed may have block sliding vertically or horizontally in mortice cut in breech. Semi-automatic QF breeches open on recoil ejecting the spent cartridge case and so facilitate rapid reloading.

Buffer See *Recoil gear*. Absorbs shock of discharge of gun slowly and so relieves strain on carriage and reduces violent backward recoil.

Burst rate of fire Of field artillery. As opposed to sustained rate of fire. A very high rate of fire for a brief period whose aim is to inflict casualties before the target can move, disperse or go to cover.

Calibration Determination of muzzle velocity of gun, which varies due to wear of bore from constant firing and other factors. Used to make guns 'shoot together' and in conformity with the map range.

Calibrating Sight Type of sight on which muzzle velocity is set on scales and applies an automatic correction to range.

Calibre Diameter of bore. The length of the barrel is sometimes expressed in multiples of the calibre.

Cannon Archaic, but sometimes still used as generic term for all 'tube' artillery.

Carriage The wheels, trail, axle, recoil, gear etc. on which the 'piece', or barrel is mounted, in a tractor drawn equipment; see also *Self-propelled.*

Cartridge Made up of charge or charges of propellant explosive contained in bags for self-sealing breech mechanisms or in fixed or semi-fixed metal cartridge cases.

Case Shot Sometimes 'canister'. Cylindrical container of bullets which opens at muzzle. Used to obtain close range lethal effect against personnel. Obsolete, but a modern version containing steel flechettes has been considered for close defence of gun positions against infantry attack.

Circular Probable Error (CPE or CEP) is defined as the circle within which 50% of the rounds will fall, the figure quoted being the radius. It is usually used in relation to missiles where the range PE and line PE are equal or nearly so.

Clinometer Used for measuring angles in vertical plane to lay for elevation.

Concentration Artillery concentration, concentration of fire. Fire of two or more batteries and on occasion of all batteries in range, brought down on one single target. Very short, intense bursts of fire from the maximum of guns appropriate to the target are the basis of current artillery tactics in the age of highly fluid armoured land warfare, and all technical development in *location*, *survey*, *calibration prediction*, *calibration meteor* and *communications* is directed to this end to make it as accurate as possible and with the shortest possible response time.

Consistency The consistency of a gun and ammunition is a measure of the dispersion of a single group of rounds about the MPI fired at a given elevation.

Counter-Battery, -Bombardment Branch of gunnery concerned with suppression of hostile artillery fire.

Correction of the moment See *Meteor.*

Covering Fire Used to suppress or 'neutralise' fire of hostile weapons during attack.

Crew Of gun. The British use 'detachment'; comprising gun commander (NCO) layer or layers, loaders, fuze-setters, ammunition handlers.

Cross Observation Means of locating and engaging targets by observation from two or more surveyed posts. See *Flash-spotting.*

Defilade Concealed by lie of land from enemy observation to the front, but able to fire freely at targets to a flank. Usually of anti-tank guns.

Defensive Fire Belt of fire or concentration of fire, usually predicted, placed in path of attack.

Dial Sight Used to measure angles in azimuth for *indirect fire.* Also called Panoramic Sight.

Direct Fire, Laying Used when the target is visible from the gun position.

Director Also (US) *aiming circle.* Angle measuring instrument resembling theodolite used for laying out lines of fire and for artillery survey.

Discarding Sabot (APDS) Composite super high velocity anti-tank shot made up of sub-calibre dart or arrow of tungsten carbide or other very hard alloy and light annular metal segments—'sabots'—of size to fit bore which are discarded on emerging from muzzle.

Driving Band Soft metal (usually copper) fitted circumferentially to projectile to engage in *rifling* and so impart spin.

Drone Originally pilotless aircraft used as target for air defence artillery. Flies a pre-set course and is equipped with camera or sensors used for surveillance and target acquisition. A remotely piloted vehicle (RPV) has a similar role but is controlled throughout its flight.

Doppler Effect Apparent variation in frequency of electromagnetic radiation emitted or reflected from a moving object as measured by a stationary observer. Used to detect movement and for the measurement of muzzle velocities. (The same phenomenon occurs acoustically; the pitch of the noise of an approaching vehicle being apparently higher and becoming lower as it recedes.)

Electronic Counter Measures (ECM) Also—Electronic Counter-Counter Measures (ECCM). Interference with the radar and radio-communications of weapons, creation of false targets etc. A complex, highly technical, highly classified activity. The general reader only requires to note that in principle any system which relies on transmissions in the electro-magnetic spectrum is vulnerable.

Elevation Angular movement of barrel in vertical plane to vary trajectory and so range. (Technical terms are 'quadrant' elevation, being sum of tangent elevation, the measure of range, and the angle of sight, being the elevation at which the gun is actually laid. The 'angle of projection' is the angle at which the projectile actually departs after aberrations due to 'jump', due to the shock of firing, and 'droop', the angular girder sag of the barrel from its point of support.)

Equilibrator Device fitted to restore balance to 'elevating mass' (i.e. barrel and recoil gear) when trunnions are not placed at centre of gravity.

Equipment (British usage) The whole weapon—gun, carriage, sights and associated gun stores such as ropes, tools etc.

Fin Stabilised Projectiles whose point-first flight attitude is determined by fins—fixed, as in mortar-bombs, or deploying immediately leaving the bore, as for some projectiles fired from smooth-bore guns. As opposed to spin imparted by rifling in the bore. (Note that some guns can fire unrotating fin-stabilised projectiles from a rifled bore, e.g. the Soviet 122 mm using a slipping driving band.)

Fire Plan Pre-planned use of artillery fire, usually based on a timed programme, if for an attack, co-ordinating all artillery fire with phases of an operation. (As opposed to engagement of targets of opportunity acquired by forward observers or surveillance devices.)

Fixed Ammunition In which the metal cartridge case is fixed to the shell or shot for ease of loading, hence 'quickfiring' for this type of gun.

Flash Spotting Method of locating guns by cross observation on flash of discharge.

Fragmentation Breaking up of the wall of an HE shell to provide fragments or splinters travelling at lethal speeds (colloquially but inaccurately 'shrapnel').

Fuze Device fitted to shell to explode HE filling on impact, or after delay so as to explode inside armour or concrete or earth emplacement, or operating by time or electronically at chosen point on trajectory.

'Grid' Arbitrary rectangular coordinate markings on map usually at 1 km spacing orientated north–south and east–west used for describing locations. See *Survey*. Basic firing data can be calculated rapidly from the grid positions of guns and targets.

Gun Used as generic term for all 'tube' artillery, also for actual *tube*, or *barrel*, or *'piece'*. Specifically and technically a long, high velocity weapon, as compared with a *howitzer*.

Gun-howitzer (Obsolete term) Weapon combining the characteristics of gun and howitzer, i.e. long range and high velocity at top charge and ability to vary charges and trajectories so as to obtain steep angles of projection and descent. In this sense most modern field artillery equipments are 'gun-hows'.

Guidance In the artillery field guided missiles include a wide variety of systems from large long range missiles used to deliver battlefield nuclear warheads to air defence missiles and anti-tank missiles. They can be divided broadly into missiles which fly and require a continuous flow of corrections and those which follow a free or ballistic trajectory receiving a *mid-course* correction, or which become fully guided in the vicinity of the target: 'Terminal Guidance'. Missiles can also be distinguished by the system employed, and the degree of automatic control. In some the correctional data are provided by an outside agency which 'tells' the missile where to go; in others the system is carried on board. (A fully automated on-board system, of course, destroys a set of expensive apparatus for every firing.) Some missiles combine two systems, e.g. using command guidance or inertial guidance to approach the target's vicinity and then changing to a homing system.

Beam rider Target is followed by a beam, e.g. laser and guidance system measures deviation of missile course from signal received.

Command guidance An outside source using a radar and computer monitors target and missile course and transmits course corrections.

Inertial guidance Based on accelerometers, as used in navigational devices. Immune to ECM.

Active Homing On-board system transmits radiation and homes on reflection from target.

Semi-Active Homer External transmitter radiates and missile homes on reflections.

Passive Homer Homes on natural radiation from the target, e.g. infra-red from hot jet efflux.

Laser Guidance Uses a laser to illuminate a target, allowing a missile to home onto the reflected radiation.

Electro-Optical Not used so far for artillery but basically observes image of the target and persists in course which keeps target picture or pattern in view.

Semi-Automatic Command to the Line of Sight Used in anti-tank guided weapons, e.g. TOW, and means that operator has only to keep sight on target and course corrections will automatically follow (as distinct from 'command guidance', in which the operator steers the missile so as to hit the target).

'Fire and Forget' Colloquial term for active or passive homing all-the-way device.

'Wire Guided' Not a guidance system but method of transferring a course correction to missile; e.g. TOW, HOT, Swingfire. The missile reels out a wire for the transmission of correctional signals which remains connected to the tracker sight.

Harassing Fire Used to interrupt movement, HQ or administrative area arrangements in rear areas, or to lower morale.

Handspikes Long, wooden, usually iron-shod levers used either fitted into sockets on trail to enable several men to lift it, or under trail to move trail laterally for big switches.

HEAT High explosive anti-tank shell, with hollow-charge HE filling.

HESH High explosive squash head anti-tank shell: Designed to open on impact placing flattened HE filling against armour plate and detonate, the subsequent shock-wave blowing off lethal fragments from inside of the armour.

High Angle Fire Using angles of projection between 45°–80°.

High Explosive (As opposed to relatively slow burning propellant explosives which can be harnessed to provide thrust for a projectile in bore of gun, or for rocket.) So called *brisant* compounds producing violent shock waves capable of shattering shell walls, damaging structures, etc. based on e.g. picric acid, tri-nitrotoluene mixed with additives.

Hollow Charge Or 'shaped' charge, usually cylindrical with concave end which on detonation focuses the explosive effect, driving jet of hot gas of great penetrating power through armour or concrete.

Howitzer Specifically a short, large calibre weapon firing a heavy shell (as compared with a gun of the same order or size) at muzzle velocities varied by increasing or decreasing the propellant charge and so at varying trajectories suited to the terrain and mission with usually steep angles of projection and of descent.

Indirect Fire From concealed gun whose layers cannot see the target; by applying firing data on the orders of a remotely sited observer or, if 'predicted', calculated in the command post.

Infra-Red Radiation from part of electromagnetic spectrum below the visible red band used for tracking some anti-tank-guided missiles, for some night vision and surveillance devices and for passive homing by some missiles.

Interdiction Loose term used to describe fire (see also harassing fire) to deny or inhibit use of area, e.g. route or defile, to enemy, or to take toll of echelons following leading assault troops.

Jacks Fitted to some field artillery carriages so as to raise wheels off the ground to promote stability and in air defence equipments also to level the carriage for firing.

Kinetic Energy Of e.g. shot or fragments, striking force due to motion, measured by multiplying mass by square of velocity at impact.

Laser Acronym from *L*ight *A*mplification by *S*timulated *E*mission of *R*adiation. Intense parallel beams of light or infra-red radiation of coherent wave-length obtained by applying energy to device called 'laser' used for guidance, range-finding. (Under development is use of laser beams to attack targets directly, but the technical obstacles are formidable.)

Laying Artillery term for aiming, having applied all data to sights. Direct laying is by sighting telescope. Indirect laying is, in azimuth, by a periscopic optical sight graduated for angular measure using an agreed aiming point or reference object and in elevation by reference to the horizontal plane supplied by spirit levels set along the line of sight and at right angles to it.

Lead Angular correction applied to sights required to hit moving target.

Limber (US 'Caisson') Two-wheeled vehicle with hook to take eye on trail and so convert gun carriage to 4-wheeled vehicle, with 'perch' and with eye to connect with hook on tractor. (Obsolete.) Carried some ready-use ammunition and gun-stores.

Location, -ing Branch of surveillance concerned with locating targets especially guns and mortars with sufficient accuracy to enable fire to be applied.

Lock Firing device, incorporated in breech mechanism. May be percussion mechanism manually operated to set off primer in metal cartridge case or tube, or electrical.

Mean Point of Impact (MPI) Centre of *zone* or *CEP*.

'Meteor' Abbr. of 'meteorological correction of the moment'. Data essential for predicted fire, e.g. air temperature and wind speeds and directions at various altitudes, barometric pressure.

Mil *See Angular Measure*.

Minelet Anti-tank, or anti-personnel delivered by shell or missile war-head. Also 'remotely delivered mines' (RDM).

Missile Used loosely for both free-flight or 'ballistic' rockets, i.e. those following a ballistic path, and also for guided weapons. E.g. SAM for surface-to-air missile, but (British) anti-tank guided weapon (ATGW) or surface-to-air guided weapon (SAGW).

Mortar Usually muzzle-loading smooth bore artillery equipment using high angle fire only. (Note: Usually mobile, light or medium equipments and simple to operate, but some are heavy and may have rifling and breech loading.)

Muzzle Brake Device using energy of gas from propellant explosion to check or mitigate recoil.

Muzzle Velocity Is the speed in m/secs or ft/secs with which the projectile leaves the muzzle, knowledge of which is essential for correct adjustment of sights to enable guns to shoot true to map and in harmony with other guns. Ascertained by measurement of wear in bore, number of equivalent full charges fired, observation of fall of shot on measured range, firing through timing screens, by photographing shell in flight and in the most modern method by electronic devices using *doppler* principle giving instant automatic read-out.

Observed Fire Indirect fire controlled by eye from a forward observation post and corrected by the fall of shot by ranging or adjusting. Given well-trained, tactically alert observers with delegated powers (so that they do not have to refer back for permission to shoot) still the most effective system for emergencies and fluid operations. It is simple and remains standard for mortar fire control.

Obturator, -ion Part of breech mechanism which prevents the escape of gas to the rear. In QF, or cartridge-sealed guns, the metal cartridge acts as an obturator. In self-sealing guns an asbestos pad is squeezed by the propellant gases to expand round the circumference of the breech face and so seals it.

Piece Archaic, but still in British use; pedantically, for 'barrel' or 'gun'.

Platform Generally, levelled or level site on which equipment is deployed in firing position. Specifically, circular metal platform on which road wheels of a towed field equipment can be placed to facilitate rapid traverse; the carriage being anchored to the platform by tie-rods and so obviating the use of trail-spades which have to be dug out laboriously once embedded if a large switch is ordered.

Predicted Fire Fire brought down by calculation without preliminary ranging. If predicted fire can be brought down accurately and quickly it has great advantages over observed adjusted fire because of its surprise effect; it is also economical of artillery ammunition. It is therefore becoming increasingly the normal method of engagement, and sophisticated equipment for range-finding, survey, meteor and computerised calculation and data storage to accomplish it is in service or under development.

Primer Initiating unit in base of cartridge which ignites propellant charge. May be fired by percussion or electric current. Also (US) the initiating element used with bag-charge guns (British 'tube').

Probable Error (PE) is defined as that distance within which 50% of the rounds will fall *or* that distance which will be exceeded as often as not.

Projectile Term embracing all shot or shell fired from a gun.

Propellant Relatively slow burning explosive used to propel shell and some solid fuel rockets. In some missiles a solid propellant is used but in some of the more sophisticated and larger ones liquid propellants, which are combinations of fuels and oxidising agents.

Ranging To ascertain range and bearing of target by observing trial shots, after obtaining rough data from map. See *Adjust*, *Register*.

Recoil Gear Spring, but more usually air/hydraulic, made up of *buffer* to absorb recoil and *recuperator* to run gun barrel out to correct firing position.

Recoilless Guns whose recoil is exactly counteracted by releasing proportion of propellant gases rearwards through vents in specially designed breech.

Recuperator Part of recoil gear which stores energy of recoil and runs barrel out to firing position ready for next round.

Register To check firing data to targets by trials shots or by survey methods in preparation for opening fire later.

Rifling Spiral grooves cut in bore of gun to grip driving band of shell and impart rotation.

Rocket, -Propulsion, -Assisted Propulsion achieved by the reaction from propellant gases vented from the rear of a missile. The propellant may be either solid or liquid. Some mortar-bombs and shells are rocket assisted; i.e. initiated from a gun and then boosted in flight by a rocket. Range can thus be extended without the need for a longer and heavier equipment, but at the cost of a rather more costly and complicated projectile, with a reduced explosive filling.

Rocking Bar Basic sighting system by which the sight bracket is tilted forward by the desired elevation and brought back to the horizontal by elevating the gun-barrel which then has the correct quadrant elevation.

Round Of ammunition, comprising projectile, cartridge and fuze, which last may be ready fixed in shell or provided separately so as to permit fuze appropriate to mission to be fitted before loading.

Saddle Part of gun carriage, on which barrel/recoil assembly rests and in turn rests on trail axle assembly.

Self-propelled Weapon system usually mounted on tracked, armoured carriage, often resembling tank but distinguished from it by virtue of the fact that it is not primarily designed for close combat or offensive use, except SP anti-tank guns. See *Assault guns*.

Shell Hollow cylindrical projectile with ogival pointed nose—container for HE, smoke canisters, minelets, bomblets or smoke producing or toxic chemicals.

Shield Steel plate fitted on the lighter towed equipments to protect crew from small arms fire and shell-fragments from front.

Shot Solid projectile for armour piercing. Either shell-shaped or a very thin lance or dart shape; made of toughened steel or very hard alloy.

Shrapnel Obsolete anti-personnel weapon consisting of lead bullets released by air-burst shell. Colloquially, lethal fragments from case of HE shell. Modern technical term is 'pre-formed, fragmented' (PFF) designed for use against either armour or personnel.

Sound Ranging Method of locating guns by recording time intervals of arrival of sound waves from shock of discharge at an array of special microphones in surveyed positions.

Spade, Trail-spade Fitted at end of trail to embed on recoil of carriage and prevent further rearward movement. See *Platform*, which also grips soil and prevents rearward movement of carriage in firing but leaves trail free to move in azimuth.

Split Function Laying In some equipments the task of laying for azimuth or 'line' and for elevation are divided between two layers.

Spotting Rifle Small calibre rifle fitted to some anti-tank equipments to obtain correct range inconspicuously.

Surveillance Used generally of devices or agencies whose role is to scan hostile territory and provide tactical and target information. See *Location*.

Survey Process of relating guns, observation posts and targets to the map grid, essential for predicted fire or for concentrating fire of separated batteries. (Note that a preliminary fix by ordinary map-reading methods by eye and compass provides an instant rough fix enabling observed fire to open from the moment of deployment.)

Switch To alter the line of fire in azimuth.

Target Acquisition Self-explanatory jargon term usually applied to detection, selection and engagement of targets; originally in air defence, now generally.

Trajectory Ballistic path of projectile. In a vacuum a parabola; in practice due to air resistance and loss of velocity it is asymmetrical, with the angle of descent much steeper than the angle of projection. (Note: the height of the vertex of the trajectory in feet is four times the square of the time of flight so that even with a field howitzer at a moderate range and a time of flight of 30 seconds this can be 3,600 feet; which emphasises the need for correct meteor data at all altitudes of ballistic interest, and also the risk to low-flying aircraft.)

Trail Long member by which gun-carriage is towed when moving. In action rests on ground forming three point base with road wheels.

Pole trail Single hollow metal tube.

Box trail Two long and two short members enabling the breech to be depressed at high elevation between the two long members.

Split trail Double trail each with spade hinged forward to open out on deployment, so enabling the gun barrel a wide angle of traverse without the need to break out the spades and move the trail and carriage.

Note: Some equipments, e.g. air defence, and the Soviet 122 mm, have instead of trails, members which on deployment provide a cruciform or other shaped base, levelled by jacks, which permit all-round traverse.

Traverse To move or 'switch' the gun relative to the carriage in azimuth—'top traverse'—or movement of the whole carriage in azimuth.

Traversing lever Metal extension at end of trail to facilitate traverse.

Trunnions Horizontal cylindrical projections from the barrel/recoil gear assembly riding in trunnion bearings permitting elevation.

Tube Synonym for piece, gun or barrel (US); tubular lining of gun permitting worn rifling to be easily changed in action; tube with firing compound used in lock of breech loading guns to initiate explosion of propellant charge.

Weapon System A comprehensive term meaning more than mere 'weapon'. The system comprises the gun, or launcher, the missile or the technical fire control system, the command system and the communications linking the various parts.

Zone Of gun. See *Probable Error*. Pattern described by large number of successive shots from gun.

BAW - P

Directory of Manufacturers of Artillery Equipment

Artillery—Towed and Self-propelled

AB Bofors Box 500, S-69020, Bofors, Sweden
Bowen-McLaughlin-York York, Pennsylvania, USA
Breda SpA Via Lunga 2, 25100 Brescia, Italy
British Manufacturing & Research Co. Ltd Springfield Road, Grantham, Lincs., England
Citefa 1603 Villa Martelli, Buenos Aires, Argentina
Commerce International Group Ave. des Arts 23, B-1040 Brussels, Belgium
Creusot-Loire 15 rue Pasquier, F-75383 Paris, France
Esperanza y Cia SA Marquina, Vizcaya, Spain
FFV Ordnance Division S-63187 Eskilstuna, Sweden
FMC Ordnance Division 1105 Coleman Ave., San Jose, CA 95108 USA
Ford-Aeronutronic Ford Road, Newport Beach, CA 92663, USA
General Dynamics Corp PO Box 2507, Pomona, CA 91766, USA
General Electric Co. Lakeside Ave., Burlington, VT 05401, USA
Giat 10 Place Georges Clemenceau, F-92211 Saint-Cloud, France
Hadfields Ltd East Hecla Works, Vulcan Road, Sheffield, S91 1TZ England
Hispano-Suiza SA Gonfreville-l'Orcher, F-76700 Harfleur, France
International Military Services Ltd 4, Abbey Orchard Street, London SW1P 2JJ England
A/S Kongsberg Vapenfabrikk N-3600 Kongsberg, Norway
Mauser-Werke Oberndorf GmbH PO Box 1349, D-7238 Oberndorf, Germany
Oerlikon-Buhrle Birchstrasse 155, Zurich, Switzerland
Oto-Melara SpA 15 via Valdilocchi, 19100 La Spezia, Italy
Pacific Car & Foundry Co. Renton, Washington, USA
Ramta Ltd PO Box 323, Beersheba, Israel
Rheinmetall International SA 100, Boulevard du Souverain, B-1170, Brussels, Belgium
Rock Island Arsenal Rock Island, Illinois, USA
Royal Ordnance Factories Directorate Northumberland House, Northumberland Avenue, London WC2N 5BP, England
Salgad 11a Catherine Wheel Yard, Little St. James' Street, London SW1A 1DS, England
Soltam Ltd PO Box 1371, Haifa, Israel
SRC International SA 114 Boulevard Brand Whitlock, B-1200 Brussels, Belgium
Oy Tampella AB Paakonttori, Lapintie 1, SF-33100 Tampella, Finland
Vickers Ltd Elswick Works, Newcastle-upon-Tyne NE99 1CP, England

Ammunition and Rockets

British Manufacturing & Research Co Ltd Springfield Road, Grantham, Lincs., England
Borletti SpA Via Washington 70, I-20146 Milan, Italy
Commerce International Group Inc. 23, Ave des Arts, B-1040 Brussels, Belgium
Diehl GmbH Fischbachstrasse 20, D-8505 Rothenbach, W. Germany
Dynamit Nobel AG Haberstrasse 2, D-5210 Troisdorf-Oberlar, W. Germany
Eurometaal NV PO Box 419, Zaandam, The Netherlands
Fabrique Mational Herstal SA B-4400 Herstal, Belgium
Ferranti Instrumentation Ltd Moston, Manchester M10 0BE, England
GIAT 10, Place Georges Clemenceau, F-92211 Saint-Cloud, France
Hirtenberger Patronenfabrik AG A-2552 Hirtenberg, Austria
IMI Kynoch (Eley) Ltd PO Box 216, Birmingham B6 7BA, England

International Military Services Ltd 4, Abbey Orchard Street, London SW1P 2JJ, England

Israeli Military Industries PO Box 7055, Tel Aviv, Israel

Luchaire SA 180 Boulevard Haussmann, F-75382 Paris, France

Mecar SA 6522 Petit-Roeulx-les-Nivelles, Belgium

Nico Pyrotechnik Postfach 1227, D-2077 Trittau, Bez Hamburg, West Germany

Olin Corp 275 Winchester Ave., New Haven, CT 06504, USA

sa PRB 168 Avenue de Tervueren, B-1150 Brussels, Belgium

A/S Raufoss Ammunisjonsfabrikker N-2831 Raufoss, Norway

Royal Ordnance Factories Directorate Northumberland House, Northumberland Avenue, London WC2N 5BP, England

Snia 162 Via Sicilia, I-00187 Rome, Italy

Sudsteirische Metallindustrie GmbH A-8430 Leibnitz, Austria

Tavaro SA 1–5 Avenue de Chatelaine, Geneva 13, Switzerland

Valmet OY PO Box 155, SF-00131 Helsinki, Finland

Wallop Industries Ltd Middle Wallop, Stockbridge, Hants, England

Fire Control Computers

Ferranti Computer Systems Ltd Western Road, Bracknell, Berks, England

Cimsa 10–12 Avenue de l'Europe, F-78140 Velizy, France

Elbit Computers Ltd PO Box 5390, Haifa, Israel

GEC Computers Ltd Elstree Way, Boreham Wood, Herts WD6 1RX, England

Communications & Radar

AEG-Telefunken AG Fachbereich N12, Elisabethstrasse 3, D-7900 Ulm, West Germany

CAP Scientific Ltd 233 High Holborn, London WC1V 7DJ, England

Contraves AG 590 Schaffhauserstrasse, CH-8052 Zurich, Switzerland

Cossor Electronics Ltd The Pinnacles, Elizabeth Way, Harlow, Essex, CM19 5BB, England

Electronique Aerospatiale PO Box 51, F-93350 Le Bourget, France

Electronique Marcel Dassault 55 quai Carnot, F-92214 Saint-Cloud, France

L. M. Ericsson S-34120 Moldnal, Sweden

Evershed & Vignoles Ltd Acton Lane, Chiswick, London W4 5HJ, England

Hewlett Packard 1507 Page Mill Road, Palo Alto, CA 94303, USA

Hollandse Signaalapparaten BV Hengelo, The Netherlands

Magnavox 1700 Magnavox Way, Fort Wayne, IN 46804, USA

Marconi Communications Ltd Marconi House, New Street, Chelmsford, Essex CM1 1PL, England

MBLE 23 rue des Courcelles, F-75362 Paris, France

Motorola PO Box 2606, Scottsdale, AZ 85252, USA

Plessey Avionics & Communications Vicarage Lane, Ilford, Essex IG1 4AQ, England

Plessey Radar Ltd Addlestone, Weybridge, Surrey KT15 2PW, England

Plessey Telecommunications Ltd Edge Lane, Liverpool L7 9NW, England

Pye TMC Ltd Swindon Road, Malmesbury, Wilts SN16 9NA, England

Racal Acoustics Ltd Wembley, Middx. HA0 1RV, England

Racal Acoustics Ltd Wembley, Middx. HA0 1RV, England

Racal Communications Ltd Western Road, Bracknell, Berks. RG12 1RG, England

Redifon Telecommunications Ltd Broomhill Road, Wandsworth, London SW18 4JQ, England

Rockwell International Cedar Rapids, IA 52406 USA

Rohde & Schwarz GmbH & Co K-G Postfach 801469, Munich 80, West Germany

Selenia Industria Elettronica SpA Via Tiburtina Km 12,400, I-00131, Rome, Italy

Siemens AG Postfach 700079, D-8000 Munich-70, West Germany.

Sperry Gyroscope Downshire Way, Bracknell, Berks RG12 1QL, England

Standard Telephones & Cables Ltd STC House, 190 Strand, London WC2R 1DU, England

Storno Ltd Frimley Road, Camberley, Surrey, England

Telemit Electronik GmbH Heidemannstrasse 17, 8-Munich-45, West Germany

Thomson-CSF 23 rue de Courcelles, F-75632 Paris, France

Simulators & Training Aids

Aero Electronics (AEL) Ltd Gatwick House, Horley, Surrey RH6 9SU, England
Ferranti Computer Systems Ltd Birdhall Lane, Cheadle Heath, Stockport SK3 0XQ, England
Invertron Simulated Systems Ltd 15 Western Road, Hurstpierpoint, Sussex BN6 9SU, England
Martin Marietta Aerospace 6801 Rockledge Drive, Bethesda, MD 20034, USA
Miltrain Ltd Middle Wallop, Stockbridge, Hants SO20 8DX, England
Redifon Simulation Ltd Gatwick Road, Crawley, Sussex RH10 2RL, England
RFD Systems Engineering Ltd Catteshall Lane, Godalming, Surrey GU6 1LH, England
SAAB-Scania AB Box 1017, S-55111 Jonkoping, Sweden
Singer Co (UK) Ltd Churchill Industrial Estate, Lancing, Sussex BN15 8UE, England
Solartron Electronic Group Ltd Farnborough, Hants GU14 7 PW, England
Tecquipment International Contracts Ltd Bonsall Street Long Eaton, Notts. NG10 2AN, England

Surveying Equipment

Barr & Stroud Ltd Anniesland, Glasgow G13 1HZ, Scotland
B & W Elektronik A/S 4 Hovmarken, DK-8520 Lystrup, Denmark
Cartographic Engineering Ltd Landford Manor, Salisbury, Wilts SP5 2EW, England
F. W. Breithaupt & Sohn PO Box 100569, 35-Kassel, West Germany
Decca Navigator Co Ltd 9 Albert Embankment, London SE1 7SW, England
Lasergage Ltd Lennig House, Masons Avenue, Croydon CR0 9XS, England
Simrad A/S PO Box 6114, Etterstad, Oslo 6, Norway
Survey & General Instruments Co. Fircroft Way, Edenbridge, Kent, England
Tellurometer (UK) Ltd Oakcroft Road, Chessington, Surrey KT9 1RQ, England

Optical & Electro-Optical Observation Instruments

AGA Infra-Red Systems Ltd Arden House, West Street, Leighton Buzzard, Beds. LU7 7ND, England
Avimo Ltd 140 Tottenham Court Road, London W1P 0JD, England
Baird Corp. 125 Middlesex Turnpike, Bedford, MA 01730, USA
Bonaventure International (Security) Ltd 18/21 Jermyn Street, London SW1Y 6HN, England
Eletronica SpA Via Tiburtina Km 13,700, Rome, Italy
Eltro GmbH Postfach 102120, D-6900 Heidelberg, West Germany
Evershed Power Optics Ltd Bridge Wharf, Chertsey, Surrey KT16 8LJ, England
MEL Ltd. Manor Royal, Crawley, West Sussex RH10 2PZ, England
Oldelft PO Box 72, 2600MD, Delft, The Netherlands
Philips USFA BV Meerenakkerweg 1, 5600MD, Eindhoven, The Netherlands
Pilkington PE Ltd Glascoed Road, St. Asaph, Clwyd LL17 0LL, Wales
Plessey Optoelectronics & Microwave Ltd. Wood Burcote Way, Towcester, Northants NN12 7JN, England
Rank Optics Ltd 200 Harehills Lane, Leeds LS8 5QS, England
Rank Precision Industries Ltd Langston Road, Debden, Loughton, Essex IG1 3TW, England
Sopelem 102 rue Chaptal, 92306 Levallois-Perret, France
Swarovski Optik K-G Absam, A-6060 Hall-in-Tirol, Austria.

Vehicles

Alvis Ltd Holyhead Road, Coventry CV5 8JH, England
Cadillac Gage Co. PO Box 1027, Warren, Mich 48090, USA
Chrysler UK Ltd Boscombe Road, Dunstable, Beds. LU5 4LX, England
Crayford Special Equipment Co. Ltd Westerham, Kent, England
DAF B/V Eindhoven, The Netherlands
Fiat Corso Marconi 10, I-10125, Turin, Italy
FMC Corp. 9400 South Dadeland Boulevard, Miami Fla 33186 USA

GKN-Sankey Ltd Hadley Castle Works, Telford, Salop TF1 4RE, England

Glover, Webb & Liversidge Ltd Hamble Lane, Hamble, Hants. SO3 5NY, England

Klockner-Humboldt-Deutz AG Cologne, West Germany

Krauss-Maffei AG Krauss-Maffei-Strasse 2, D-8000 Munich 50, W. Germany

Laird (Anglesey) Ltd Beaumaris, Anglesey, Gwynedd LL58 8HY, Wales

Man K-G Nuremberg, West Germany

Magirus-Deutz AG PO Box 2740, D-7900 Ulm, West Germany

Marshall of Cambridge Ltd Airport Works, Cambridge CB5 8RX, England

Renault Vehicules Industrielles 33 quai Gallieni, F-92513, Suresnes, France

SCM Panhard et Levassor 18 Ave d'Ivry, F-75621 Paris, France

Steyr-Daimler-Puch AG Karntner Ring 7, A-1010 Vienna, Austria

Sibmas Rue Henri Pauwels 85, B-1400 Nivelles, Belgium

Stonefield Vehicles Ltd Caponacre Industrial Estate, Cumnock, Ayrshire, Scotland

Vauxhall Motors Ltd PO Box 3, Luton, Beds. LU2 0SY, England

Volkswagen GmbH, Wolfsburg, West Germany

Volvo International S-40508 Goteborg, Sweden

Wyvern Trading (Exeter) Ltd Berkeley House, Dix's Field, Exeter, Devon, England

Index of Weapons

The reader experienced in weaponry who wishes to refer to the characteristics of a weapon of known identity, or whose function or approximate size he knows, can most conveniently trace its particulars by using the contents list. He can then consult the appropriate section in the text bearing in mind that the weapons are arranged in ascending order of calibre or missile diameter: e.g., all towed 57 mm anti-tank guns are together, the earlier marks preceding the later. If only the code-name e.g., Gainful; acronyn, e.g. Amets, or initials, e.g. TOW is known, these are in alphabetical order in the first part of the index. Weapons and equipments described by calibre, mark or date, e.g. 152 mm howitzer M 1938 (M 10) are under sections of the index corresponding to the sections in the text and in the same order as the contents list; not alphabetical. E.g. 105 mm SP gun Abbot —FV 433 can be found in the alphabetical index under 'Abbot', and under '105 mm gun' in the self-propelled field gun section.

Field Guns and Howitzers

Self-propelled Field Guns

Mortars

Artillery Rockets

Ancillary Equipment for Surface-to-Surface Artillery

Anti-tank Guns and Guided Weapons

Anti-aircraft Guns and Surface-to-air Missiles

Air-Defence Ancillary Equipment

Coast Artillery